T0344943

The Path to 5G in the Developing World

SUSTAINABLE INFRASTRUCTURE SERIES

The Path to 5G in the Developing World

PLANNING AHEAD FOR A SMOOTH TRANSITION

 **WORLD BANK GROUP**

Contents

Boxes

Figures

Photo

Tables

Foreword

Fifth-generation (5G) mobile technologies have the potential to unlock a new wave of digital transformation, products, and services across the globe. Providing significantly higher internet speeds, reduced latency, and an enhanced capacity for simultaneous connections, 5G improves connectivity for individuals and organizations and offers diverse applications across sectors. According to estimates by the GSM Association, 5G will add as much as $950 billion to the global economy by 2030, with an increasing share from developing countries toward the end of this decade. 5G rollout is well under way, with nearly 300 operators in more than 100 countries launching commercial 5G services by January 2024, contributing to an estimated 1.5 billion connections worldwide as of the end of 2023.

Through its improvements in speed and performance, 5G can be a catalyst for societal and economywide change and advancing the United Nations Sustainable Development Goals. 5G can support productivity gains across various sectors relevant to developing countries—from revolutionizing agriculture with precision techniques to reshaping urban living and sustainability through intelligent transport systems and smart grids to bridging the gap to quality education for all by supporting real-time virtual learning. By helping drive global goals in health, education, energy, and infrastructure, 5G can contribute to creating more sustainable, inclusive, and resilient economies and societies.

Governments have a strategic role to play in unlocking 5G's potential as well as in simultaneously ensuring equitable access to meaningful connectivity for all. Policy makers can enhance the conditions for 5G deployment and adoption, including by managing spectrum proactively and providing regulatory flexibility and frameworks to support network deployment not only for 5G but also for telecommunications networks more generally. In addition, policy makers must ensure rural and lower-income communities are not overlooked, as these areas will continue to rely on earlier-generation networks and devices for the near future. To that end, it is important to remain technology neutral and instead focus on the outcome of achieving universal, high-quality, and affordable broadband access. Finally, governments can safeguard against the potential risks of 5G and other mobile network technologies, including by ensuring environmental sustainability and protecting against cybersecurity threats.

As 5G's rollout and adoption continue to grow, efforts must focus on leveraging these technologies' strengths to build a more connected, resilient, and sustainable world while minimizing risks. This report offers policy makers a detailed understanding of the opportunities for, challenges and risks of, and strategies to establish a thriving ecosystem for 5G's deployment and adoption in the developing world. The World Bank is dedicated to fostering a global digital environment that is accessible, secure, and forward looking, ensuring that the benefits of 5G extend to every corner of an increasingly interconnected world.

Guangzhe Chen
Vice President for Infrastructure
World Bank

Acknowledgments

The preparation of this book was led by Rami Amin and Doyle Gallegos, under the leadership of Christine Qiang and Casey Torgusson of the Digital Development Global Practice at the World Bank. The core report team members are Niccolo Comini, Natalija Gelvanovska, Kyoung Yang Kim, Hyea Won Lee, and Maria Claudia Pachon from the Digital Development Global Practice; Tania Priscilla Begazo Gomez from Finance, Competitiveness, and Innovation Global Practice; Vivien Foster from the Imperial College London and formerly chief economist for infrastructure at the World Bank; Edward Oughton, assistant professor at George Mason University; Je Myung Ryu of the Government of the Republic of Korea and formerly of the World Bank's Digital Development Global Practice; and Zhijun William Zhang, technology and innovation adviser at the Bank for International Settlements. Boutheina Guermazi and Mark Williams provided valuable guidance during the book's earlier stages.

The team is grateful to the many reviewers who provided thoughtful insights and guidance at various stages of the report's preparation. External peer reviewers include Dror Fixler, professor and research scientist at Bar Ilan University; William Lehr, research scientist, Computer Science and Artificial Intelligence Laboratory, Massachusetts Institute of Technology; Stephen Unger, former chief technology officer and head of strategy at the UK Office of Communications at Ofcom; and Adam Weinberg, chief technology officer, FirstPoint Mobile Guard. Internal peer reviewers include Cecilia Briceño-Garmendia, Yusuf Karacaoglu, Juan Navas-Sabater, Rajendra Singh, and a team at the International Finance Corporation consisting of Ariana Batori, Georges Vivien Houngbonon, Charlotte Kaheru, Julian Ernesto Angus McNeill, Carlo Maria Rossotto, Davide Strusani, and Ferdinand Van Ingen.

Clara Stinshoff coordinated and managed the publication of the report. Kelly Alderson and Breen Byrnes led the communication and dissemination efforts, with input from Topaz Mukulu. Patricia Katayama led acquisitions, and Christina Davis managed production. Sandra Gain copyedited and

Ann O'Malley proofread the report. William Pragluski designed the cover. Hadiza Eneche provided the team with resource management and administrative support.

This report was supported by the Digital Development Partnership, which aims to advance digital transformation in low-income and middle-income countries by building strong digital foundations and accelerators and facilitating use cases for the digital economy to thrive.

Executive Summary

INTRODUCTION

Deployments of fifth-generation (5G) mobile network technologies are taking off globally, with more than one billion connections reached by the end of 2022. This deployment includes 5G trials, licensing, and network deployments. Interest among governments in understanding the implications of this technology is rising as the numbers of base stations, use cases, and 5G subscribers continue to grow.

This book provides an overview of the technical capabilities of 5G, discusses how they may impact the information and communication technology (ICT) industry, and explores how 5G can support the Fourth Industrial Revolution (Industry 4.0) and help countries make progress toward the United Nations Sustainable Development Goals (SDGs). The book also guides policy makers to better understand the opportunities, challenges, and risks posed by 5G, as well as steps that can be taken to establish an enabling ecosystem for 5G deployment and adoption.

5G IS A SIGNIFICANT STEP FORWARD IN MOBILE BROADBAND

The three main capabilities of 5G mobile technologies as compared with fourth-generation (4G) mobile network technologies are much higher speeds, lower latency (lag delay in data transmission), and enhanced capacity for connecting large numbers of users simultaneously with minimal interference. 5G's flexible design enables technologies to be adapted to different circumstances and needs in both public and private campus network settings. While maximum capabilities have yet to be fully achieved and commercialized, target capabilities include peak data rates of 20 gigabits per second, transmission latency of 1 millisecond, connection density of 1 million devices per square kilometer, and 100 times more network energy efficiency per data traffic as compared with 4G.

The key underlying technologies that enable this functionality include multiband spectrum usage, which can support different use cases; massive multiple input, multiple output technology and beamforming, which enable suppression of interference and higher spectral and power efficiency; mobile edge computing, which enables ultra-low latency; network virtualization, which allows for quicker system updates by virtualizing many functions; and network slicing, which enables one physical 5G network to operate as several customized virtual networks offering different services and performance characteristics to suit various purposes. Some of these technologies are revolutionary; others enhance or scale existing 4G-enabling technologies. The deployment model of 5G—whether as a standalone deployment with a 5G core or a non-standalone deployment that integrates 4G core—also has implications for the availability and performance capabilities of these technologies.

5G WILL SUPPORT DIGITAL TRANSFORMATION AND CAN HELP DEVELOPING COUNTRIES MEET THE SDGs

5G represents a significant shift in the role that mobile networks can play in society. It builds on the capabilities of 4G Long-Term Evolution through improvements in the speed and performance of which 5G networks are comparatively capable. The increased emphasis on industrial and commercial applications in 5G could also enhance the productivity and performance of business and public sector organizations, particularly in developing countries where alternative forms of high-speed broadband connectivity or ways to connect high-density devices in different sectors are often limited.

The potential of 5G is still being explored by mobile operators and industrial users. These applications will develop over time as networks are deployed and users find new ways to integrate them into their processes. It is too early for reliable assessments of the economic value of 5G, because most use cases have not yet been fully tested—and are rarely being tested in developing country settings. However, the following are current areas of investigation that have potential for development:

- *Precision agriculture*, which entails real-time, micro-targeted optimization of inputs such as water, fertilizer, and pesticides to enhance farm yields, improve sustainability, and reduce waste along agricultural supply chains.

- *Smart cities*, which enable more sophisticated management of the urban space; closer integration of urban systems and services; and better management of perennial urban challenges such as sprawling urban development, congestion, pollution, and crime.

- *Intelligent transport systems*, which permit more proactive management of multimodal transport networks and services and growing reliance on autonomous vehicles, all of which can improve road safety as well as the use of passenger time and road space.

- *Smart grids*, which support the transition to a more flexible, decentralized, and renewable-based energy system while enabling a more sophisticated management of energy demand and increasing the efficiency and efficacy of infrastructure maintenance.

Some of these applications are being developed using earlier generations of mobile technologies, but the enhanced functionality of 5G creates further opportunities for more sophisticated solutions. In some contexts, the optimal approach will be one based entirely on 5G networks. In others, it will be a combination of 5G and earlier generations. Although peculiar aspects of 5G's technology and performance capabilities resulting from features like network slicing have generated much excitement, new use cases with 5G are difficult to predict, and the use cases that may unfold cannot be foreseen. As with earlier generations of mobile network technologies, commercial use cases drive transformational change, as was seen in the case of third-generation (3G) mobile network technologies and subsequent developments with handsets and applications.

5G NETWORK DEPLOYMENT WILL START IN URBAN AREAS AND IS LIKELY TO COEXIST WITH OTHER TECHNOLOGIES FOR MANY YEARS

As with the previous generations of mobile technologies, 5G is capital intensive and will require network operators to invest large amounts in deploying the new networks if 5G is to deliver on the technologies' target performance capabilities. This high level of investment is driven by the need for more base stations than with previous mobile generations, increased fiber backhaul to mobile sites, and a shift to multi-access edge computing, among other requirements. Despite these drivers pushing costs upward, some factors, such as network softwarization and infrastructure sharing, will reduce costs.

The economics of 5G mean that most deployments will initially be focused in urban areas, which are more commercially attractive than rural areas, particularly in developing countries. Industrial campuses will also be early adopters of private 5G networks, deployed alongside (or replacing) other modes of network connectivity. Applications of 5G networks in these campus settings could be a key driver of network deployments outside core urban areas.

The speed of the deployment and the extent of 5G's penetration into rural areas will vary by country. 5G fixed wireless access is one option for last mile solutions in areas where fiber or copper is limited, but its viability and value proposition will depend on a variety of contextual factors.

In middle-income countries and those where the population is more geographically concentrated, the coverage of 5G networks could extend further, including into semi-urban and rural areas. This expanded coverage would open opportunities for deeper digital transformation in both the public and private sectors.

5G's HIGH DEPLOYMENT AND ADOPTION COSTS CAN BE REDUCED

The cost of 5G network equipment—as well as of handsets and user devices—is expected to decrease over time, as has been the case for previous generations of network technologies. Despite this reality, deploying a 5G network will remain expensive for developing countries; the cost has generated debate among policy makers and other stakeholders on its value proposition over 4G networks. The deployment costs can be reduced in various ways, including the following:

- *Addressing spectrum costs.* High-spectrum license costs can negatively affect the profitability of network businesses overall and may reduce incentives to deploy networks in financially marginal areas, which are often in rural settings. Governments must consider the trade-offs between increased revenue generated through spectrum sales and the potential implications for coverage and pricing of mobile services. Policy makers can improve spectrum efficiency by supporting innovative methods of spectrum sharing for both licensed and unlicensed spectrum, as well as by addressing the adequacy of available spectrum to support 5G's performance capabilities. Clarifying pricing and criteria for reallocating bands and issuing technology-neutral spectrum licenses will help telecom operators transition from one service to another, thereby expediting 5G deployment.

- *Upgrading network backhaul.* The capacity of network backhaul—the links that connect mobile sites to the rest of the network—is a key constraint on overall network performance. A majority of mobile sites in developing countries use wireless microwave technologies for backhaul. These technologies are cost-effective and flexible, but they are also subject to limitations on network performance. Fiber-optic links are needed for backhaul if 5G is to meet its performance potential. The high costs of these technologies could be reduced through concerted efforts by governments and network operators, such as by leveraging cross-sector infrastructure sharing with the energy and transport sectors and the availability of right-of-way access.

- *Sharing public infrastructure.* The costs of establishing mobile sites are a major component of the overall cost of building a 5G network, largely due to network densification. These costs could be reduced if governments made public land, buildings, roadside infrastructure, electricity transmission towers, and other types of infrastructure available to network operators.

- *Enabling and encouraging network sharing.* Regulatory frameworks that encourage network sharing can substantially reduce the costs of 5G.

For many consumers in developing countries, the cost of 5G handsets is likely to be prohibitive for adoption in the early years of network deployment. Although the experience of handsets for previous generations of mobile technologies indicates that handset prices fall over time, 5G-enabled handset affordability will remain a challenge for low-income households for the near future. In most industrial and public sector applications of 5G, device affordability for end users is a less significant constraint on adoption.

5G POTENTIALLY INCREASES SOME TYPES OF RISK THAT MUST BE MANAGED BY GOVERNMENTS AND REGULATORS

Cybersecurity risks, health concerns, and environmental risks may erode public trust and confidence in 5G networks. If they are not properly addressed or managed, these concerns could become an obstacle for 5G adoption.

- *Addressing cybersecurity risks.* The wider and deeper application of broadband in business and government applications using 5G networks also means that the consequences of cyberattacks on 5G networks could be much more far-reaching as compared with earlier generations of mobile networks. Despite the significant advances in establishing a more secure network architecture through intentional "security by design" features, 5G introduces new intrinsic vulnerabilities such as software flaws and supply chain risks. The increased data flows that 5G applications will yield will also increase the surface area vulnerable to various forms of cyberattacks. For countries with low cybersecurity capacity, the risk will be amplified considerably. It is expected that cybersecurity threats will continue to evolve as 5G networks and use cases are commercialized. Risks may also be amplified by the coexistence of 5G with less secure mobile generation networks such as 4G. This issue underscores the importance of governments setting and enforcing minimum security standards and increasing domestic capacity for managing cybersecurity risks.

- *Managing environmental risks.* Production and consumption of 5G technology may increase the carbon footprint of mobile networks and ICT systems more broadly if the increase in data consumption and device and equipment use exceeds the significant efficiency gains anticipated from 5G's target technology capabilities. The increased data traffic over 5G networks will require significant data center capacity; consequently, the efficiency and sustainability of these data centers will play an important role in minimizing 5G's overall carbon footprint. The associated accumulation of e-waste from antiquated mobile equipment and internet of things devices will further compound environmental harm. Despite 5G being designed according to stringent energy efficiency standards, especially when compared with previous generations of mobile networks, initial deployments have evidenced an upsurge in overall energy consumption due to the wider variety of applications. This issue raises important concerns about climate change for governments and industry. As of 2022, further research is needed for accurate measurement and forecasting of the net impact of 5G networks.

- *Investigating health issues.* Although public concerns about health risks from exposure to electromagnetic fields associated with 5G and other mobile technologies can be significant in some countries, concrete evidence directly linking the use of wireless devices to general health issues is lacking. As research continues, governments should actively update their guidelines on exposure to electromagnetic fields and enforce compliance based on the latest guidance and evidence produced by external bodies such as the International Commission on Non-Ionizing Radiation Protection, the International Telecommunication Union, the World Health Organization, and others. Even in the absence of any scientific evidence linking 5G networks to adverse health consequences, governments must address public concerns and manage communications if disruption is to be avoided. This work includes providing access to the latest evidence and research that has been published by the aforementioned organizations and other trusted institutions with expertise in this area.

DEPLOYMENT OF 5G NETWORKS CAN BE FACILITATED THROUGH GOOD POLICY

Governments can enhance their country's capacity to deploy and adopt 5G in several policy areas, including strategic policy coordination, innovation ecosystems, competition policy, spectrum management, and regulatory frameworks.

- *Define an integrated national vision for 5G.* Countries should define a collective, strategic vision for how 5G can advance the national development agenda. Policy makers should devise their national 5G strategies to encapsulate assessments of their country's position on each of the key building blocks of 5G, notably, the existing coverage of high-speed mobile broadband (4G), fiber-optic backhaul, spectrum policy, regulatory frameworks, and whole-of-government collaboration. The latter is important in the case of 5G to ensure that the policy implications, which affect a wide range of economic and social activities beyond the ICT sector, are considered in an integrated fashion. A national vision for 5G should also include a focus on demand-side initiatives, in consultation with the private sector and other stakeholders.

- *Strengthen the innovation ecosystem.* Where relevant to their economies, governments should encourage industry to explore, innovate, and develop business models that enable monetization of a 5G-enabled economy. Late 3G and early 4G adoption were driven by innovations in application software that earlier technologies could not support. Therefore, more innovation and focus on 5G-enabled services and applications that are not possible with 4G are needed, accompanied by policy design to strengthen the innovation ecosystem that can create an enabling environment to spur this exploration. 5G trials and dedicated testbeds and labs can be helpful tools because they focus on the entire process of business development and not just on the technology itself. Parallel experimentation is needed on the policy side through regulatory sandboxes and trial spectrum licenses, which can provide the private sector with the flexibility to innovate and the public sector with the opportunity to learn how to adapt the regulatory environment. As with a national strategic vision, the development of a supportive innovation ecosystem for 5G should also focus on demand-side initiatives in consultation with private sector and other stakeholders.

- *Consider optimal market structure.* In recent years, some countries have seen consolidation in the mobile market. One justification presented for this consolidation has been to allow operators to finance the capital investment required to compete. The investment implications of 5G have prompted renewed calls to allow further consolidation in some markets. The case for this is complex, and countries must consider carefully how to balance competition and investment incentives among mobile network operators in light of both the new competitive dynamics generated by 5G as well as the enhanced opportunities for significant cost savings from increased infrastructure sharing along with more flexible approaches to spectrum and licensing.

- *Release spectrum early.* Regulators should enhance institutional capacity to secure and release enough spectrum, including globally harmonized pioneer bands, while avoiding the risk of spectrum fragmentation that prevents 5G from delivering on design performance. Given that spectrum allocation will not only be of interest to traditional telecom operators but may also be relevant for industry verticals operating private networks, regulators must strategically balance competing demands for spectrum from new and incumbent users. The design of spectrum assignment methods, spectrum pricing, and spectrum licensing regimes all have a material impact on the viability of 5G networks and associated investment incentives. Regulators should be mindful of the transparency of spectrum assignments and the affordability of spectrum fees. Because 5G non-standalone deployment leverages existing 4G infrastructure, a technology-neutral approach to spectrum licensing is important, and allowing licensees to re-farm spectrum to use it in the most efficient way could achieve significant gains. To address increasing data traffic in 5G, spectrum authorities also should pay attention to the role of unlicensed technologies, such as next-generation Wi-Fi, and to balance the use of licensed and unlicensed spectrum in the spectrum management framework. Carrier aggregation within the same technology in different frequency bands and across various technologies will also need a flexible and forward-looking approach.

- *Introduce regulatory flexibility.* Providing greater flexibility for network deployment through regulatory accommodations is critical for reducing costs and improving the viability of network deployment. Facilitating access to passive infrastructure, such as buildings and streetlights, for sites through a supportive policy and regulatory framework will help ease network deployment. In addition, governments should improve the regulatory environment to support and encourage backhaul investment and infrastructure sharing among operators.

Policy makers should focus on setting the enabling environment early, even while investment plans remain under deliberation, particularly given that most of the best practices that can ease the path to becoming a 5G nation also reflect best practices for general telecommunications policy and regulation.

As the next generation of mobile network technologies, 5G is a significant step forward in terms of network capabilities. Its enhanced performance and ability to be integrated into industrial and commercial processes provide new opportunities and use cases for businesses and governments. However, as with previous generations of mobile technologies, the deployment of 5G infrastructure in the developing world may proceed more slowly due to structural impediments. These issues include the relatively low density of demand, the high costs of base stations and antennas, the limited availability of fiber backhaul, and the prohibitive costs of 5G-enabled handsets and associated data packages, as well as broader digital development challenges around digital skills, digital trust, and locally relevant content. Nevertheless, governments can do much through policy and regulatory measures to significantly reduce deployment costs, mitigate potential risks, and ease the path to a 5G future.

Abbreviations

1G	first-generation mobile network technologies
2G	second-generation mobile network technologies
3G	third-generation mobile network technologies
3GPP	3rd Generation Partnership Project
4G	fourth-generation mobile network technologies
5G	fifth-generation mobile network technologies
5G-PPP	5G Infrastructure Public Private Partnership
64T64R	64 transmitting and 64 receiving antenna elements
ACS	American Cancer Society
AI	artificial intelligence
ARENA	Australian Renewable Energy Agency
B2B	business-to-business
B2C	business-to-consumer
Capex	capital expenditures
CENELEC	European Committee for Electrotechnical Standardization
CIS	Commonwealth of Independent States
CISA	Cybersecurity and Infrastructure Security Agency (United States)
DSS	dynamic spectrum sharing
EMF	electromagnetic field
ENISA	European Union Agency for Cybersecurity
EPA	Environmental Protection Agency (United States)
ETSI	European Telecommunications Standards Institute
EU	European Union
FCC	Federal Communications Commission (United States)
FWA	fixed wireless access
GSA	Global Mobile Suppliers Association
IARC	International Agency for Research on Cancer
ICNIRP	International Commission on Non-Ionizing Radiation Protection

ICT	information and communication technology
IEEE	Institute of Electrical and Electronics Engineers
IMSI	international mobile subscriber identity
IMT	international mobile telecommunications
IoT	internet of things
IPCC	Intergovernmental Panel on Climate Change
ITU	International Telecommunication Union
ITU-R	International Telecommunication Union Radiocommunications Sector
ITU-T	International Telecommunication Union Telecommunications Standardization Sector
LoRa	long range
LTE	Long-Term Evolution
M2M	machine-to-machine
MENA	Middle East and North Africa
mMIMO	massive multiple input, multiple output
mmWave	millimeter wave
MNO	mobile network operator
NB-IoT	NarrowBand–internet of things
NCC	Nigerian Communications Commission
NICT	National Institute of Information and Communications Technology (Japan)
NIST	National Institute of Standards and Technology (United States)
OECD	Organisation for Economic Co-operation and Development
ONF	Open Network Foundation
RAN	radio access network
RBS	rogue base station
RF	radiofrequency
SDGs	Sustainable Development Goals
UNCTAD	United Nations Conference on Trade and Development
WHO	World Health Organization
WIA	Wireless Infrastructure Association
WRC	World Radiocommunication Conference

UNITS OF MEASUREMENT

EHz	exahertz
Gbit	gigabit
Gbps	gigabits per second
GHz	gigahertz
h	hour
Hz	hertz
KHz	kilohertz

km	kilometer
m	meter
Mbit	megabit
MHz	megahertz
ms	millisecond
PHz	petahertz
s	second
THz	terahertz
TWh	terawatt hours

1

Overview of 5G Technology: Capabilities and Global Adoption Trends

KEY MESSAGES

- The three main capabilities of fifth-generation (5G) mobile network technologies as compared with fourth-generation (4G) mobile network technologies are much faster speeds, lower latency, and enhanced capacity for simultaneously connecting large numbers of users.

- 5G's key design principles are flexibility and modularity. 5G can be adapted to the circumstances and needs of consumers, businesses, government, and both public and private campus network settings.

- 5G will coexist with earlier generations of mobile technology for some time, the extent and duration of this coexistence will vary between countries, and its development can be encouraged where commercially feasible.

- Although 5G trials and pilots are being tested in most countries, the majority of commercial 5G networks and adoption is in high-income countries as of 2023.

- In many developing countries, the initial adopters of 5G are likely to be in industrial sectors with specialized requirements, such as manufacturing, before wider adoption by consumers.

- Variance in 5G deployment and adoption may contribute to an expanded digital divide between high- and low-income countries and between urban and rural areas—particularly in the medium term. However, the practical effect of the divide may not be as severe as with previous digital divides and can be mitigated with alternative modes of high-speed mobile connectivity to achieve broader development outcomes.

INTRODUCTION

5G mobile network technologies have been the subject of intense public interest long before they left the proof-of-concept stage. These technologies have the potential to deliver data up to 20 times faster than 4G Long-Term Evolution (LTE), with a nearly imperceptible delay (also known as "latency"), while simultaneously connecting large numbers of devices. Given its high performance capabilities, 5G is prompting governments and policy makers to think differently about the role that mobile networks play in society—particularly when integrated with other Industry 4.0 technologies, such as artificial intelligence (AI), cloud services, and edge computing—and how they should strategically prepare their countries for the changes ahead.

Mobile technologies have radically reshaped modern life and catalyzed economic growth around the world. The first generation of cellular mobile communications debuted in the 1980s, with each subsequent generation manifesting about every 10 years. Second-generation (2G) mobile network technologies added narrowband data services in the mid-1990s, making it possible to carry limited amounts of data. With the development of the Global System for Mobile Communications standard, third-generation (3G) mobile network technologies expanded to wideband communications and inaugurated videoconferencing. Along the way, handsets evolved into smartphones, and 4G became a priority, offering better data handling for media, applications, and communications. The arrival of 5G holds even greater potential, with a peak data rate under ideal conditions set at 20 gigabits per second and a target latency rate of 1 millisecond (ITU 2015).[1]

The transition to 5G will be challenging for all countries—especially those with developing economies. The cost associated with 5G site architecture will make it difficult to attract investments for deployment, especially in rural areas, unless incentives are offered. The lower average revenue per user in these settings further reduces the attractiveness of the investment. For mobile network operators, extending the existing 4G LTE infrastructure—and integrating it efficiently with 5G deployments—will be difficult and expensive. This issue is partly due to the cost of upgrading networks, but it also reflects significant innovations in the design of mobile networks. These challenges are expected to be magnified in developing countries, where networks can be more expensive to operate.

That said, when 5G is viewed in the context of the sectors where it has the greatest potential—manufacturing and private business campuses—the affordability equation changes, especially under favorable spectrum assignments. 5G's modularity permits improvements in mobile data handling to be applied directly to business requirements, offering the potential for substantial gains in productivity.

This publication explores the latest innovations in mobile technology, the opportunities created by 5G across key sectors, fundamental risks around deployment, and policy and regulatory options to address some of these challenges. A key objective is to help policy makers in developing countries understand how they can plan ahead to ease the introduction of 5G and maximize the potential value it can provide, given national development objectives.

This introductory chapter sets the stage by describing 5G technologies, standards, enablers, and trends. Chapter 2 examines 5G's potential in greater detail, focusing on Industry 4.0 and key sectors. Chapter 3 analyzes the risks associated with 5G networks, focusing on cybersecurity, sustainability, and health concerns. Chapter 4 concludes with a discussion of policy and regulatory challenges and recommendations across priority aspects, including spectrum management, regulatory frameworks, market structures, business models, fiscal regimes, institutional governance, and capacity building.

Although this book provides a comprehensive overview of what 5G mobile network technologies could mean for developing countries, it is by no means an exhaustive review of 5G's impact in every aspect of a country's economy. The goal is not to advocate for any specific technology, ranging from other mobile generations such as 4G LTE to the evolving wireless 802.11 standards (the Wi-Fi family of local network systems that usually reach up to 100 feet)—all of which have unique advantages, depending on circumstances and technical requirements. Instead, the book's objective is to provide a balanced overview of the significant opportunities that 5G offers, some key risks, and how government decisions can ease the path to deployment.

A key motivation for this publication was to respond to growing inquiries from governments about the rapid development of 5G technology and how to prepare for future deployment. This book fills a gap in the ongoing discourse by focusing more on the implications of 5G for developing countries, which vary by digital maturity. Much of the literature to date skews toward 5G's economic benefits and sectoral opportunities in developed countries with more mature industries, with less attention paid to alternative connectivity options or the challenging circumstances faced by governments in developing countries, which may necessitate a different approach to policy design. Therefore, this book focuses on options for policy makers in developing countries that would maximize the gains from enhanced 5G connectivity while also facilitating network connectivity more broadly, regardless of technology choice or deployment timing.

Some findings and forecasts presented in this book are time limited due to the rapidly evolving nature of 5G technologies and associated supply chains, the economics of deployment, innovations with new business models, and evolving geopolitical considerations. Readers are encouraged to use this book as a preliminary guide to key topics for low- and middle-income countries, along with other reports that provide deeper dives into the evolving landscape.

5G DESIGN PRINCIPLES, CAPABILITIES, AND ENABLING TECHNOLOGIES

The definition of 5G is derived from the International Telecommunication Union (ITU), which endorses the development of standards for all generations of cellular network systems. 5G is also called "IMT-2020," with "IMT" signifying international mobile telecommunications and "2020" being the anticipated year of commercial launch, following earlier ITU goals such as IMT-2000 for 3G.[2] Each generation of mobile technology standards is defined by a series of vision statements and performance requirements to which countries agree through the ITU. The 3rd Generation Partnership Project (3GPP)—an independent consortium of seven organizations devoted to the development of telecommunications standards—then transforms those agreements into detailed technical standards for the development of products and services, which the ITU then considers for endorsement.[3]

5G is still evolving, and its improvements are identified by release numbers. 3GPP Standards Release 15 marked the first phase of 5G's introduction in 2018, defining the fundamental technologies for radio access nodes and core network functions. These standards enabled the first wave of 5G products to be designed and promoted. Release 16 was published in July 2020, and Release 17 was published at the end of 2022.[4] These releases extend existing capabilities with new network management tools and add features addressing the needs of new user groups and deployment scenarios, such as the industrial uses of the internet of things (IoT), operations in unlicensed spectrum, and intelligent transportation systems (Peisa et al. 2020).

From 4G to 5G: Design Parameters, Key Improvements, and Deployment Strategies

In its IMT-2020 vision statement, the ITU identified eight parameters that give 5G its unique profile and enable a large set of use cases (ITU 2015) (refer to table 1.1).[5] As of 2023, the commercial 5G networks that have been deployed remain far from their maximum capabilities across all targets, and so maximum performance capabilities should not be expected. At the same time, trade-offs exist between peak data rates and latency.

5G is designed to accommodate continuing growth in mobile data traffic and to provide a user experience with high speed and responsiveness while at the same time exceeding fixed networks in accessibility and versatility. Its key design principles are flexibility and modularity, which allow it to be applied in many use cases and scenarios. "Modularity" refers to 5G's capacity to integrate various combinations of "radio, baseband, and cloud technologies flexibly across multiple spectrums and distributed architectures" (Headrick 2019). Thus, 5G can be adapted by circumstances and needs, making it promising for business, industry, and consumers given the changes in user demands, which include higher expectations for instantaneous connectivity; a shift in emphasis from human-to-human communications to machine-to-machine

TABLE 1.1 **Key target capabilities of 4G and 5G**

Capability	Description	4G	5G
Peak data rate	Maximum achievable data rate under ideal conditions per user or device	1 Gbit/s	20 Gbit/s
User-experienced data rate	Data rate available to mobile users and devices throughout the coverage area	10 Mbit/s	100 Mbit/s
Latency	Radio network's contribution to the time elapsed between the source sending a packet and the destination receiving it	10 ms	1 ms
Mobility	Maximum speed at which a defined quality of service and seamless transfer between radio nodes belonging to different layers or radio access technologies can be achieved	350 km/h	500 km/h
Connection density	Total number of connected or accessible devices per unit of area	10^5 devices/km^2	10^6 devices/km^2
Network energy efficiency	For networks, "energy efficiency" refers to the quantity of information bits transmitted to and received from users per unit of energy consumed by the radio access network		100 times more than 4G
Spectrum efficiency	Average data throughput per unit of spectrum bandwidth, per cell		3 times more than 4G
Area traffic capacity	Total traffic throughput per geographic area	1 Mbit/m^2	10 Mbit/m^2

Source: ITU 2015.
Note: 4G = fourth-generation mobile network technologies; 5G = fifth-generation mobile network technologies; Gbit = gigabit; h = hour; km = kilometer; m = meter; Mbit = megabit; ms = millisecond; s = second.

communications; and support for exponential growth in data traffic, high user density, and high mobility communications (for example, crowded express trains, congested city centers, or thronged sporting events).

The first version of 5G, which was the basis of most commercial launches as of 2022, was engineered to operate alongside an existing 4G LTE network upgraded to support 5G devices. The 5G core network is also designed to work within a heterogeneous network and be accessible via 5G New Radio, 4G, Wi-Fi, or the fixed wireless broadband network (GSM Association 2019d). These critical features can help accelerate 5G adoption while also enabling multiple strategies for deployment based on existing systems and economic considerations.

Although 5G has the capacity to outperform 4G on every parameter, the improvements that likely will have the most impact are speed and latency. As shown in table 1.1, a 5G user can attain a peak data transfer rate that approaches 20 billion bits per second under ideal conditions (which requires being close to the base station, among others)—20 times faster than that possible with 4G LTE. This improvement is due to broader bandwidth, advanced antenna technology, and 5G's threefold gain in spectrum efficiency, as measured by the amount of information that can be transferred over a given bandwidth (ITU 2015). The peak values in table 1.1 represent theoretical maximums, rather than commercially available or even possible thresholds, given the near-perfect conditions that would be required to achieve them.

Consequently, these figures should be used for comparison with similar theoretical peaks set for 4G.

Latency, which is also called "lag" or "delay," measures the time it takes for a data packet to reach its destination and for a response to be sent back, facilitated by the radio network (the time it takes for a "ping" to traverse the radio interface). 5G commercial networks will have a noticeable impact on user experience and enable new applications across sectors and use cases where the least possible latency is needed, and these effects will be even more acute where mobile computing is enabled at the edge of networks, closer to the end users. The ITU groups dozens of 5G use cases into three general usage scenarios—enhanced mobile broadband; ultra-reliable, low-latency communications or mission-critical, machine-type communications; and massive machine-type communications or massive IoT—as presented in table 1.2.

Migrating from 4G to 5G does not require turning off and replacing an old network with a new one. 5G allows a range of deployment strategies, including gradual transitions that can be tailored to the needs of specific operators in specific locations in specific markets. Thus, the transition from 4G LTE to 5G includes the option of using an upgraded 4RepuG core network to support connectivity with new 5G radio and 5G-enabled user equipment. This option is known as "non-standalone." Most early commercial 5G launches have taken this path.

TABLE 1.2 **Proposed 5G usage scenarios and use case examples**

ITU usage scenario	Description	Use case examples
Enhanced mobile broadband[a]	Mobile broadband addresses the human-centric use cases for access to multimedia content, services, and data. The enhanced mobile broadband scenario entails new application areas and requirements in addition to improved performance and a seamless user experience for existing mobile broadband applications.	• Enhanced indoor and outdoor broadband • Enterprise collaboration • Augmented and virtual reality
Ultra-reliable, low-latency communications or mission-critical, machine-type communications	This use case has stringent requirements for throughput, latency, and availability.	• Autonomous vehicles • Smart grids • Remote patient monitoring and telehealth • Industrial automation
Massive machine-type communications or massive IoT	Characterized by many connected devices typically transmitting a relatively low volume of non-delay-sensitive data. Devices are required to be low cost and have a long battery life.	• IoT • Asset tracking • Smart agriculture • Smart cities • Energy monitoring • Smart home • Remote monitoring

Source: ITU 2015.
Note: IoT = internet of things.
a. IHS Markit (2019) and GSM Association (2019d) consider fixed wireless access an enhanced mobile broadband use case.

Alternatively, 4G components can be replaced with a 5G core network and 5G New Radio to create a "standalone" 5G network that offers the full range of 5G performance improvements that more closely approach peak capabilities. Realistically, standalone is likely to be a long-term process, extending over the next decade in developing countries, as many countries are still constructively using earlier mobile technologies to support development objectives.

Key Enabling Technologies for 5G Networks

Among the many technologies that account for 5G's unique functionality and performance, the following are the most significant:

- Multiband spectrum usage, which includes the high-band space above 24 gigahertz (GHz), also referred to as the "millimeter wave (mmWave) spectrum"

- Multiple input, multiple output; massive multiple input, multiple output (mMIMO); and beamforming antennas

- Mobile edge computing

- Network softwarization

- Transmission, backhaul, and xHaul technologies.

Some of these technologies are evolutionary; others enhance or scale existing 4G-enabling technologies. The technologies support non-standalone 5G and standalone 5G systems, as described later in this chapter.

- *Non-standalone 5G.* Approved in December 2017 as part of 3GPP Release 15, the non-standalone 5G architecture introduces a new 5G radio access network (RAN) alongside upgrades to existing 4G networks. Its capacity for dual connectivity supports links to 5G user equipment with 4G and 5G New Radio systems. Combining a 4G core with a 5G RAN can open new frequency ranges and use techniques such as mMIMO, beamforming, and carrier aggregation.

  Signals from both 4G and 5G New Radio can be combined to increase usable bandwidth (DeTomasi 2018), which raises spectral efficiency, increases network capacity, and boosts performance by leveraging already deployed 4G LTE networks and base stations. So far, most countries launching 5G have used the non-standalone configuration.

- *Standalone 5G.* Approved in June 2018 as part of 3GPP Release 15, the standalone architecture is an end-to-end 5G network that can operate independently of a 4G core network and is not "backward compatible." When fully standardized, instead of relying on dedicated hardware, the standalone core will enable cloud-native and service-based architecture to support network slicing, software-defined networking, network virtualization, and mobile edge computing.

The technological attributes of the five 5G-enabling technologies are presented in the following section, and relevant policy and regulatory implications are presented in chapter 4.

Multiband Spectrum Usage and mmWave

Radio frequencies are needed to connect the users of a network with an access point. 5G networks are currently being allocated bandwidth under 3GPP standards across a range of frequency bands. Frequencies below 1 GHz constitute the low band, as noted in table 1.3. The mid band consists of frequencies between 1.5 and 7.12 GHz, and the high band consists of frequencies above 7.12 GHz.

So far, the focus for 5G introduction has been on the mid band (especially 3.5 GHz), as it offers a good combination of coverage and capacity. Although the high band enables mmWave (for example, 26 GHz or 28 GHz), which is essential for realizing the high data transfer rates of 5G, its coverage areas are small. The different spectral characteristics of the three bands correspond to different usage scenarios and raise different policy questions, as summarized in box 1.1.

TABLE 1.3 Frequency bands internationally designated for IMT identification

Band group	Spectrum band	IMT identification
Low band (below 1 GHz)	• 614–698 MHz • 694–790 MHz	614–698 MHz was identified for IMT in some region 2 countries at WRC-2015. 694–790 MHz was allocated to mobile services at WRC-2012 and became effective after WRC-2015.
Mid bands (1.5–7.12 GHz)	• 1,427–1,518 MHz • 3,300–3,400 MHz (33 African countries, 6 Asian countries, and 6 Latin American countries) • 3,400–3,600 MHz (not all countries in Region 3) • 3,600–3,700 MHz (Canada, Colombia, Costa Rica, and the United States) • 4,800–4,990 MHz (worldwide on a primary basis, although not adopted by all countries)	Global IMT identification occurred at WRC-2015. The WRC-23 agenda includes a review of the 6 GHz band and whether it should be allocated to mobile for IMT identification.
High bands (above 7.12 GHz)	• 24.25–27.5 GHz • 37–40.5 GHz • 40.5–42.5 GHz • 42.5–43.5 GHz • 45.5–47 GHz (in 53 countries, mostly in Africa but including Brazil, the Republic of Korea, and Sweden) • 47.2–48.2 GHz (in Region 2, plus 69 countries in Regions 1 and 3, including Australia, India, Japan, and the Republic of Korea) • 66–71 GHz (potential use for multiple gigabit wireless systems on a license-exempt basis)	Global IMT identification occurred at WRC-19 (with some restrictions on land mobile service use for the 24.25–27.5 GHz, 37–43.5 GHz, and 47.2–48.2 GHz bands). 45.5–47 GHz and 66–71 GHz bands can also be used for aeronautical and maritime mobile services.

Sources: GSM Association 2021b; Moura Gomes 2021.
Note: WRCs are held every few years to update the International Radio Regulations. 614–698 MHz is not identified for IMT in Region 1 (Africa, the Commonwealth of Independent States, Europe, the Middle East west of the Persian Gulf, including Iraq and Mongolia) or Region 3 (the Federated States of Micronesia, the Solomon Islands, Tuvalu, and Vanuatu in the frequency band 470–698 MHz, and Bangladesh, Maldives, and New Zealand in the frequency band 610–698 MHz). GHz = gigahertz; IMT = international mobile telecommunications; MHz = megahertz; WRC = World Radiocommunication Conference.

BOX 1.1

5G frequency ranges, spectrum bands, and usage scenarios

Low Band

Because most early fifth-generation (5G) mobile network technologies are anchored in existing fourth-generation (4G) mobile network technologies infrastructure, they can use frequencies designated for 4G if the mobile network operator and regulator agree. Low-band 5G has channel widths similar to 4G. The increased data speed is due to 5G's greater spectral efficiency, influenced by antenna characteristics and distance from the base station. 5G should normally be at least 20 percent faster than 4G Long-Term Evolution (LTE) in the same bandwidths. The lower the frequency, the longer distances signals can travel and the better the coverage. Therefore, base stations can be more widely separated, making for lower-cost coverage extending beyond urban and suburban areas into rural areas. Low-band signals also penetrate walls to serve indoor users.

Operators can offer 5G technology using LTE frequencies over low-band spectrum using dynamic spectrum sharing, which intelligently switches frequency channels between 5G and LTE depending on the number of users or changes in traffic. This innovation has been used by operators (such as AT&T and Verizon in the United States) to expand 5G coverage rapidly (Fletcher 2020).

Mid Band

Mid band includes frequencies in the 1.5–7.125 gigahertz (GHz) range, which is emerging as the favored choice among commercial telecom operators and policy makers. 5G can deliver more data in this frequency range than in the lower range. Although 5G can be licensed for nationwide coverage, a few countries, including Germany,[a] have offered private companies the exclusive use of a portion of the band.

High Band

High band provides 5G's fastest data speeds, as spectrum assignments above 7.125 GHz can carry more data. However, the waves travel only a short distance, and walls, trees, and people can block them. The high band's limited signal range is a problem for serving thinly populated areas, and easy blocking compromises deployments in cities. Telecom operators must invest in massive multiple input, multiple output and beamforming technologies to make up for the fragile nature of the waves in this band. Thus, the private sector's appetite for developing uses in this band may be low. Governments can promote innovation by increasing the demand for use cases through research and development, public-private partnerships, and foreign direct investment. Cases exist in which the limitations of the high band are not a problem, such as on business campuses and for localized applications within a custom network.

a. For Germany's provision details, refer to https://www.bundesnetzagentur.de/DE/Sachgebiete /Telekommunikation/Unternehmen_Institutionen/Frequenze.

5G requires a mix of low, mid, and high bands to support all use cases. Table 1.4 summarizes key 5G performance variables for the bands in the spectrum.

In the 5G era, policy makers and regulators must strengthen their capacity for spectrum management to make larger contiguous blocks of spectrum available and improve network performance for consumers. With increasing demand for limited spectrum from private companies, policy makers should maximize the social value of spectrum allocations over maximizing revenues. Policy makers can also allow innovative methods of spectrum sharing for both licensed and unlicensed spectrum to improve spectrum efficiency. Providing trial spectrum bands to encourage innovation will help enterprises to explore new 5G use cases. Clarifying pricing and criteria for reallocating bands and issuing technology-neutral spectrum licenses will help telecom operators transition from one service to another, thereby expediting 5G deployment.

mMIMO and Beamforming

mMIMO can integrate many antennas on a radio panel. This innovation can enable suppression of multi-user interference between data streams and achieve spectral efficiency (Lu et al. 2014). Arrays of up to 64 transmitting and 64 receiving antenna elements (64T64R) are being developed for 5G (Chataut and Akl 2020). These large arrays achieve high spectral efficiency and higher data capacity but require sophisticated hardware and software for signal processing and two to three times more electric power than 4G (Clark 2019). Photo 1.1 illustrates a 5G base station with a 64T64R mMIMO radio with 64 antenna elements.

mMIMO enables beamforming, which in turn minimizes interference, a key benefit of 5G over previous mobile network technologies. Using active antenna arrays with associated signal processing enables the base station to

TABLE 1.4 **Summary of performance variables for 5G spectrum**

Variable	Low band 5G	Mid band 5G	mmWave 5G
Typical 5G bands (ITU region dependent)	600 MHz 700 MHz 800 MHz	2.6 GHz 3.3 GHz, 3.5 GHz 3.7 GHz, 4 GHz	24 GHz, 26 GHz, 28 GHz 39 GHz, 40 GHz 47 GHz, 50 GHz
Typical spectrum assignments	2x10 MHz	80–100 MHz recommended In EU, often lower	1,000 MHz recommended 100 MHz in FCC 103 auction
Relative speeds	20% faster than LTE	6x faster than LTE	10x faster than LTE
Coverage (typical cell radius)	Good 2–4 km	Average 0.5–2 km	Poor 100–500 m
Economics (cost to cover an area)	Very good	Average	Very poor
Antenna size	Large	Smaller	Individually small, collectively large

Source: Behar 2019; Remmert 2022.
Note: 5G = fifth-generation mobile network technologies; EU = European Union; FCC = US Federal Communications Commission; GHz = gigahertz; ITU = International Telecommunication Union; km = kilometer; LTE = Long-Term Evolution; m = meter; MHz = megahertz; mmWave = millimeter wave (the high-band space above 24.25 GHz).

PHOTO 1.1 5G base station with a 64T64R mMIMO radio antenna array

Source: Hardesty 2019.
Note: In the photo, the plastic cover of the array has been removed. 5G = fifth-generation mobile network technologies; 64T64R = 64 transmitting and 64 receiving antenna elements; mMIMO = massive multiple input, multiple output.

transmit slightly different variations of the same signal, which constructively interferes at a point in space away from the antenna. This concentrated signal toward a specific device refers to "beamforming."

In urban environments, signals may ricochet off nearby buildings and reach their target indirectly, which may cause an unintended echo effect, with multiple copies of the same signal being received. mMIMO receivers account for this by using signal-processing algorithms to combine data from all return paths (correcting for differences in path length) to improve the quality of the received signal.[6] The result is a tightly focused link, or beam, between a sender and receiver with less power wasted, less interference, and better support for more communicators. Figure 1.1 illustrates how mMIMO and beamforming technologies are used together to provide optimal coverage of areas using 5G networks.

Concerns about the potential health effects of 5G increased significantly during the COVID-19 pandemic, with mMIMO and beamforming attributes of the technology sometimes cited as potential risks. Although no scientific evidence confirms the fears about the adverse health effects of 5G radiofrequency waves, governments should strengthen institutional capacity to develop, monitor, and implement relevant public health standards. In addition, regulators should consider harmonizing power density limits in accordance with international guidelines, to allow more flexibility for telecom operators in 5G deployment. mMIMO and beamforming technologies will increase the energy consumption of 5G networks over 4G.

FIGURE 1.1 Active antenna systems and mMIMO

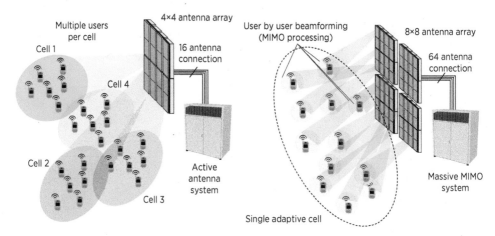

Source: Newsome, Parekh, and Matharu 2018.
Note: mMIMO = massive multiple input, multiple output.

MOBILE EDGE COMPUTING

5G's ultra-low latency is further supported by mobile edge computing (although 5G standards do not presume the availability of this network architecture). In conventional cellular networks, communications to and from subscribers are processed in a data center beyond the 5G network. Mobile edge computing reduces the need to send everything to and from the core. Some processing can occur close to the subscriber, which reduces the time needed to process and route data through the network and shrinks the bandwidth needed to link mobile sites with the core network, a process known as "backhaul." Mobile edge computing can also be used to store and run the radio software that formerly was embedded in each base station.

On the one hand, depending on the number, size, and complexity of edge computing centers, this site architecture may be more costly to implement than pre-5G designs. This issue may lead operators to rent "edge cloud" capacity from third parties (Taleb et al. 2017) to bring computing and storage to the edge of the local network. On the other hand, mobile edge computing may yield greater savings due to the reduced need for bandwidth in the network backbone. The economics are contingent on the details of the configuration.

NETWORK SOFTWARIZATION

Changing or adding a new function to a network traditionally requires modifying or replacing hardware. However, an ongoing trend with 5G is to separate hardware from software, which can reduce replacement of hardware where software updates can suffice. By implementing functions using software, it is possible to update the network without site visits from technicians. This approach also reduces the amount of specialized hardware required by networks, which could reduce capital investment in instances in which capital expenditure for hardware replacement is greater than the operating expenditure needed to support software integration.[7]

Although traditional hardware-software solutions may have superior technical performance over "off-the-shelf" hardware, innovations within the industry are advancing rapidly toward supporting this decoupling. Key benefits include reduced vendor lock-in with hardware and reduced barriers to market entry among new vendors of software and generic hardware. However, this new dynamic does not preclude the risk of vendor lock-in with software.

Earlier cellular generations had projects like OpenAirInterface, OpenBTS, OpenCellular, OpenLTE, and srsLTE, which enabled low-cost, community-owned networks to develop in areas lacking commercial service (Song 2016). These earlier projects typically were small, whereas current 5G projects are backed by large corporations and governments. Furthermore, 5G standards are designed to encourage the disaggregation and virtualization of functions, whereas in earlier generations, experimenters had to overcome the standards' lack of support for such strategies. The major suppliers supply virtualized networks, which are vertically integrated between the software and hardware components. While virtualization results in various forms of efficiency, due to, for example, increased operational flexibility, it does not imply the more complicated disaggregation of networks, which is becoming more practical as open RAN standards develop.

As discussed in chapter 3, the characteristics of softwarization increase the vulnerability of 5G networks to coding errors, covert introduction of "back doors," and cyberattacks. However, this approach has many advantages, including faster implementation of changes; greater transparency, flexibility, and configurability; and potentially large cost savings.

Beyond the goal of implementing the network as generic, off-the-shelf, vendor-neutral hardware, network softwarization enables cloud networking. Through cloud networking, users have access to hyperscale cloud providers without having any hardware other than radio units. This approach not only lowers the cost of initial network deployment but also provides access to AI technologies embedded in cloud companies.

In addition, network slicing enables one physical 5G network to operate as several customized virtual networks offering different services and performance characteristics to different users on the same platform. Network slicing enables new forms of wholesale access, supporting a greater degree of service differentiation.

Threats exist to software-defined networking and network virtualization and slicing. Relevant mitigation measures should be detailed in the cybersecurity paradigm. In addition, regulators must take a balanced approach to net neutrality in the context of network slicing so as not to hinder innovation.

TRANSMISSION, BACKHAUL, AND xHAUL TECHNOLOGIES

The final critical enabling technology for 5G is related to the technologies that carry data between parts of the network, referred to as "transmission," "backhaul," and "xHaul." When data packets for communications travel by radio from the handset to the base station, they have only begun the journey. These data packets move from the base station to their final destinations through what is traditionally referred to as a "backhaul network," which links a mobile operator's RAN with the network core. These links can involve various materials and technologies, microwave, optical fiber, and satellite transmissions.

Point-to-point microwave is a common form of backhaul that, in some cases, can be used for transmission. Microwave installs quickly and does not require excavations that disrupt street traffic. Globally, a vast majority of all mobile backhaul is currently implemented with microwave, although the percentage varies significantly by region (Lombardi 2019).

The upfront installation costs of microwave are lower than those of optical fiber. However, microwave networks consume more power than fiber networks, which must be considered when evaluating the lifetime costs of various technologies. Moreover, microwave networks have lower capacity than fiber-optic networks.[8] As mobile data traffic has increased in volume, operators have increasingly deployed fiber in their backhaul and transmission networks. Unfortunately, installing fiber is time consuming and costly, and permits may be required to stop traffic and dig up streets. Therefore, microwave deployment in low-income countries and rural areas has been limited, although this situation may change.

Satellites provide another long-distance backhaul option. Traditionally, this option has utilized a high-altitude orbit, which creates significant latency owing to the distance signals must travel. Projects are under way to launch small, low-orbit satellites to support mobile broadband. Low-orbit satellites move across the sky quickly, so large numbers are needed to prevent the loss of connectivity (Tikhvinskiy and Koval 2020). Since IMT Release 14, 3GPP has been developing standards for low-orbit satellite backhaul in areas without adequate terrestrial infrastructure (3GPP 2018).

The question of which medium is more cost effective for backhaul depends on the time frame, volume of data traffic, and physically available options where base stations are located. Looking ahead, the GSM Association estimates that by 2027, the use of fiber and satellite for backhaul will continue to grow globally as a percentage of total backhaul, while copper connections will continue to decrease, reflecting a recent trend, as illustrated in figure 1.2.

Table 1.5 summarizes the current backhaul options presented in this section, although copper is listed only as a reference point because it is not a realistic consideration for supporting the 5G network backhaul.

Given that backhaul capacity is a key determinant of 5G network performance, policies accelerating the buildout of backhaul that is scaled proportionally to network performance are critical. As optical fiber is the preferred 5G backhaul, especially in areas of high data traffic, priority should be given to establishing a fiber-ready environment, for example, by providing a favorable environment for investment and promoting the use of so-called "dark fibers" (pre-installed in anticipation of future use).

FIGURE 1.2 **GSM Association forecasts for global backhaul connections, 2025 and 2027**

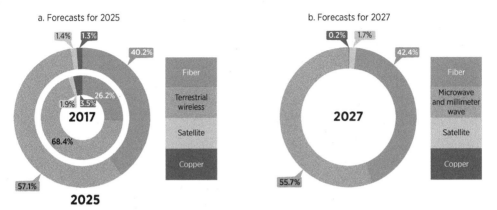

Sources: Data for 2017 and forecasts for 2025 are from Camargos (2019); forecasts for 2027 are from GSM Association (2021c).

TABLE 1.5 **Current alternatives for backhaul media**

Dimension	Copper wire	Optical fiber	Microwave	Satellite
Future usage trends for 5G networks	Limited usage for 5G due to low data capacity and pricing increases linearly with bandwidth	This optimal choice is expected to account for 40 percent of backhaul worldwide by 2025	The most widely used technology currently requires access to spectrum	WRC-2019 adopted measures to protect satellite services in bands adjacent to IMT-2020. Further studies on 5G access to satellites are in 3GPP Release 17.
Deployment costs	Low	High, although costs are expected to decrease	Low	Medium. High-density fixed satellite service costs are decreasing.
Operational costs	Low indoors, high outdoors	Low	High	High
Meets 5G standards	No	Yes	Yes, V-band (60 GHz) and E-band (70–80 GHz) can deliver 10–25 Gbps and satisfy 5G standards. Bands under 40 GHz can also meet 5G requirements and are expected to dominate backhaul globally according to the GSM Association.	Partially. Latency is an issue. For geostationary satellites, network delay ranges from 500 to 600 ms; low-orbit satellites have latencies of 50 ms.
Suitable for remote areas	No	No, not cost effective for remote areas	Yes	Yes. It can also perform complementary roles with other technologies, for example, as a temporary network in disaster recovery areas.

Sources: GSM Association 2018; Viasat 2019.
Note: 5G = fifth-generation mobile network technologies; Gbps = gigabits per second; GHz = gigahertz; IMT = international mobile telecommunications; ms = millisecond; WRC-2019 = World Radiocommunication Conference 2019.

Whereas in traditional mobile networks backhaul is the transport link between the RAN and the core network, "fronthaul" refers to the transport link between a base station's baseband unit and the radio unit. In the 5G era, baseband units have been further disaggregated into a remote unit and a central unit, with the transport link between the two referred to as "midhaul," as illustrated in figure 1.3.

"xHaul" refers to the combination of backhaul, midhaul, and fronthaul transport links interconnected as part of a unified network. This architectural shift was developed to support 5G networks, which will have more small cells and towers than ever before. The disaggregated xHaul links enable the flexibility to support new and different use cases with 5G that have different requirements and other efficiency advantages.[9]

Regardless of which deployment strategy is adopted, equipment for infrastructure inputs—such as broadband access, microwave and optical transport, mobile cores, and RANs, along with associated routers and switches—is limited to a highly concentrated group of suppliers. The 5G equipment market is dominated by five vendors: Ericsson, Huawei, Nokia, Samsung, and ZTE (Lee 2020) Fluctuations in market structure and players can also affect network deployment costs.

COSTS AND INVESTMENT CHALLENGES

Delivering the significant infrastructure upgrades that 4G and 5G require due to their requisite investment costs can be challenging for developing countries given the much lower average revenues per user as compared with segments in high-income countries. Fluctuations in revenues determine the time needed to recover investment costs and can also affect stock prices.

FIGURE 1.3 **Disaggregated fronthaul, midhaul, and backhaul links**

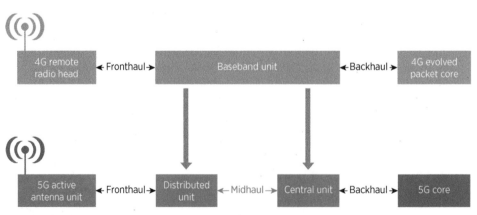

Source: Johns Hopkins Applied Physics Laboratory 2021.
Note: 4G = fourth-generation mobile network technologies; 5G = fifth-generation mobile network technologies.

Over the past decade, the average revenue per user has been falling for cellular networks around the world. In both high-income and upper-middle-income countries, revenue shrank by approximately 3 percent annually over the decade leading to 2020, to US$33 for high-income countries and US$13 for upper-middle-income countries (GSM Association 2020c). Similar rates of decline have been observed in lower-middle-income countries (3 percent annually) and low-income countries (2 percent annually), leading to 2020 per user figures for these two cohorts of US$11 and US$6, respectively. This finding suggests a significant obstacle for new mobile network infrastructure investment over the coming decade, particularly 5G deployment.

Consequently, accelerating the upgrade and expansion of mobile networks in many developing countries calls for strategic policy decisions that improve the viability and sustainability of commercial services, thereby reducing the demand for scarce state resources. As policy makers and other stakeholders in developing countries begin to consider the feasibility of 5G network deployment, they must consider the main cost drivers that determine the speed and cost of widening access. The extent of past sunk investments in fiber and towers has a significant impact on the viability of new cellular technologies, as does competition and cooperation in broadband infrastructure development. Although competition is generally desirable to create a dynamic and efficient market, demand may not be high enough in remote geographic areas to support more than one infrastructure provider, and infrastructure sharing facilitated by supportive government regulations may enable network expansion at significantly lower cost (Oughton et al. 2022). Supportive policy and regulatory actions—particularly for spectrum licensing—are explored further in chapter 4.

GLOBAL 5G DEPLOYMENT AND ADOPTION STATUS

Although it was originally seen as a successor to 4G LTE, 5G will coexist with 4G, 3G, and 2G for the near future, particularly in developing countries, partly because of the many pre-5G handsets still in use and partly because of the difficulty of expanding 5G coverage in some areas due to the short signal range of the frequencies most used by 5G.[10] The GSM Association predicts that by 2025, one-third of the world's population will have 5G coverage, and 1.2 billion connections will be on 5G networks.[11]

As of June 2022, 496 operators in 150 countries and territories had invested in 5G networks, which included tests, pilots, license acquisitions, planned deployments, and actual deployments.[12] Trials and tests are critical steps for easing the path to 5G deployment, because they provide the necessary feedback to regulators to assess how conducive the regulatory environment is for 5G deployment, adoption, and usage by different categories of users in different settings.

The vast majority of 5G deployments in emerging markets thus far have used the 3.5 GHz band, also known as "mid band" or "C-band 5G." This band can serve urban, high-density areas, but it is less capable of penetrating inside buildings than lower-frequency bands. Initial 5G deployments and service offerings are mainly in urban areas and mainly for fixed wireless access (FWA) to enhanced mobile broadband, as discussed in box 1.2. This issue may change as more 5G-capable handsets enter the market.

Although 5G FWA has the potential to accelerate connectivity with competitive advantages in cost, performance, and deployment speed, it is not

---

BOX 1.2

Will 5G fixed wireless access advance or remain as a supplementary option in broadband markets?

Fifth-generation (5G) mobile network technologies fixed wireless access (FWA) has enjoyed considerable commercial popularity in the first batch of 5G deployments. Almost half of the mobile network operators that launched 5G commercial services used FWA as their first 5G use case. Device trends also indicate that 5G FWA customer premise equipment is the most prevalent type of 5G device after 5G phones.

Global 5G FWA growth is accelerating and increasingly contributing to home broadband connections in developing markets, where the broadband penetration rate is low. Increased demand for FWA solutions to fill the gap in broadband networks has garnered further attention due to the widespread explosion of remote work and e-learning using home broadband bandwidth, underscoring the need for more robust networks. For example, the Philippines is using FWA as a last mile solution for national broadband service, and the Kingdom of Saudi Arabia is focusing on delivering FWA to rural and suburban areas where fiber or copper is not available. 5G FWA has the potential to deliver last mile connectivity to broadband infrastructure in rural areas in which fiber is not an option.

What Drives 5G FWA's Commercial Popularity?

The value proposition of 5G FWA can be split into three types of market demand: temporary, urban, and rural:

- *Temporary demand.* The COVID-19 pandemic highlighted the importance of broadband access for both businesses and homes.
- *Urban demand.* Whether replacing aging wireline infrastructure or building anew, 5G FWA can help control costs and reduce complexity in urban areas.
- *Rural demand.* 5G FWA can reach rural outskirts without major construction work, yielding a reasonable return on investment.

(continued)

BOX 1.2 *(continued)*

Government backing is essential for FWA to be adopted. Consequently, some governments are raising funds for 5G FWA to cover rural areas. For example, the Connect America Fund in the United States supports broadband initiatives in underserved communities, and the Connecting Europe Broadband Fund performs a similar role in the European Union.

No One Solution Fits All

Because 5G FWA has substitutes, the market opportunity will vary with technology, the commercial model, and local conditions (Wilson 2020). Broadband provision involves trade-offs among throughput speeds, latency, coverage, and cost. Although this issue depends primarily on the infrastructure footprint of a given operator, partnerships to secure wholesale access to backbone networks and specialized technologies will increasingly become a pragmatic consideration in the costs of 5G and fiber deployments (GSM Association 2019a).

The Developmental Context

In underserved urban areas in developing countries where physical wire infrastructure is limited—a feature many developing countries face—the possibility of using FWA 5G technology to bring high-performance 5G connectivity to customers has garnered significant attention. In addition to underserved urban areas, areas with low population density, particularly at the edges of urban areas and more rural areas where installing fiber cables is expensive, FWA also represents a viable alternative. Together, these two scenarios are found extensively across developing country contexts.

The scope of the opportunity for fixed operators will be determined not only by demography and topography but also by the cost and availability of infrastructure for deploying fiber (Wood 2019). In recent years, FWA has become a part of national broadband programs in Australia, Canada, France, Indonesia, Myanmar, New Zealand, Nigeria, the Philippines, the Kingdom of Saudi Arabia, South Africa, Sri Lanka, and Trinidad and Tobago.

a universal application. A broadband network is highly dependent on factors such as population density, subscriber uptake, average revenue per user, customer premise equipment, device cost and availability, network contention, and labor costs. However, given the competitive advantages and strong need for better broadband connection and services, FWA may often be the most cost-effective solution for high-speed broadband services for homes and businesses in developing markets. However, because FWA uses significantly more capacity per home than a mobile subscriber, mmWave will be needed to make FWA viable in the long term.

5G AND IMPLICATIONS FOR THE DIGITAL DIVIDE

The "digital divide," which is a matter of definition and degree, refers to an issue that has preoccupied policy makers since the 1970s (Van Dijk 2006). This divide highlights the differences in the availability and use of information and communications technologies between countries and regions within countries and is increasingly used to describe differences across demographics and geographies within regions. Policy makers understandably want to avoid contributing further to widening disparities.

The goalposts in the debate over the digital divide have shifted over time. In the pre-internet era, the main concern was access to voice communication infrastructures. With the advent of the internet, the emphasis shifted to digital connectivity. As with previous generations, the uneven distribution of 5G networks and adoption rates has significant potential to widen the digital divide—particularly if and when 5G's target capabilities are achieved on a commercial scale—further underscoring the need to ensure inclusiveness.

Bandwidth inequality is closely linked to economic capacities and fluctuates with technological progress and diffusion (Hilbert 2016). Hence, a development question that arises with every iteration of technology is whether the new technology will widen the digital divide. If measured by the traditional indicator of population penetration, the advent of 5G is likely to widen the digital divide between high- and low-income countries and between urban and rural areas, just as 4G did. This issue may be exacerbated for 5G because some of its deployments will be on the high-frequency spectrum, which has a much shorter range than lower-frequency bands. At the national level, high-density, high-income countries will be better able than poorer, less dense countries to deploy 5G.

However, the digital divide will be meaningful only if the adoption of broadband services that require high speeds is widespread. That is, the difference in user experience between 5G and 4G will have to be significant before the digital divide can be said to be widening in a meaningful way. With the 5G applications that require high capacity and low latency on a citywide basis, the difference will be between having such services or not. In this case, a regional divide will reemerge, as the investments required for 5G will confine such services to affluent urban areas, at least in the immediate future.

For campus network applications, which are localized and, in many cases, private, the digital divide is less relevant. Companies and strategic business considerations will determine where such campuses will be built, and thus the locational constraints will be the fiber backhaul and backbone links, which may rule out locations distant from major cities in low- and middle-income countries. As such, it is not the 5G infrastructure that is a constraint but rather the overall availability of fiber backbone networks. In addition, given the anticipated value proposition of 5G connectivity for factories and the manufacturing sector, as well as for urban smart cities where there will be high demand, countries and regions that have a limited factory and manufacturing footprint will be less likely to see 5G deployment and adoption as compared to countries and regions where factories and advanced manufacturing hold a more prominent place in the local economy.

Governments can support 5G trials and testbeds to gain a better understanding of applications for local use cases and to reinforce the innovation ecosystem. Chapter 4 further discusses the imperative of policy support for modern production operations using 5G through the provision of sufficient backbone and high-frequency spectrum in key locations.

In the longer term, mass market applications of 5G and reductions in handset prices and data costs to end users will drive an expansion of 5G coverage beyond campus-like clusters. Deficits in coverage may significantly delay the introduction of such applications in emerging countries while also preventing businesses in these countries from leveraging 5G-enabled opportunities. Chapter 4 offers recommendations for addressing these issues—even through robust 4G network connectivity in the near to mid-term—in a 5G-enabled economy.

The availability of affordable handsets and lower entry-level subscription prices will also be key determinants of the digital divide in the 5G era. The record in countries that were among the first to deploy 5G networks does reflect a decrease in average selling prices. It is difficult to predict whether this figure will reach an affordable level in developing countries within this decade, given that 3G- and 4G-capable phones and service packages remain out of reach for many.

Other long-standing barriers to digital inclusion and mobile internet adoption will mediate the speed and trajectory of 5G adoption in developing countries. For example, with data-intense applications, digital literacy and cybersecurity awareness will become significantly more important in the 5G era and must be considered when assessing the full scope of the digital divide.

The first few years of the 5G era so far have demonstrated that a growing divide is evident in both deployment and adoption patterns across countries and regions. This issue has been exacerbated in areas already facing barriers to digital inclusion with older mobile technology. Without special attention paid to developing countries and regions, most of the benefits from a 5G-enabled global economy will continue to accrue mainly to the few already well-connected countries and users, and disproportionally toward private industry, which is explored further in chapter 2.

NOTES

1. The ITU specifications for 5G reflect a trade-off between peak-data rates and latency.
2. For a brief history of the development and deployment of IMT technologies, refer to ITU (2015).
3. For more details on 3GPP, refer to https://www.3gpp.org/about-3gpp.
4. For details on Release 16 for 5G, refer to https://spectrum.ieee.org/tech-talk /telecom/standards/5g-release-16. For details on Release 17, refer to https:// www.3gpp.org/specifications-technologies/releases/release-17.

5. Recommendation ITU-R M.2083-0; refer to https://www.itu.int/dms_pubrec /itu-r/rec/m/R-REC-M.2083-0-201509-I!!PDF-E.pdf.

6. For a more detailed explanation of mMIMO, refer to https://www.avnet. com/wps/portal/abacus/solutions/markets/communications/5g-solutions /understanding-massive-mimo-technology/.

7. Theoretically, this is true only if the hardware is sufficiently advanced to accommodate all future requirements, scalable to meet all future loads, and adaptable to run any software. Rakuten, Japan's largest mobile virtual network operator that is building the first end-to-end virtualized cloud-native network, uses six different hardware platforms. Whereas softwarization may involve the use of open-source software in building and managing 5G networks, the software in an "open architecture" need not be open source.

8. The maximum theoretical throughput for a single-core fiber has been calculated at 1.2 petabits per second, with low path loss (Essiambre et al. 2010). One petabit per second was achieved in 2020 using a single-core multimode optical fiber (NICT 2020). One petabit is the equivalent of 1 million gigabits, so the speed is thousands of times faster than any backhaul alternatives.

9. While most of this book refers to backhaul links and investments, more information about xHaul—an increasingly developing architecture to support 5G—is available in DHS (2021).

10. "The 4G RAN is about 10 times denser than the 3G network, and that densification is predicted to continue through 2022 before new 5G equipment takes over the growth trend. It is predicted that 5G networks will need to be 10 times denser than 4G networks, a 100-fold increase over 3G" (Getto 2019).

11. Refer to GSM Association (2023).

12. NTS Statistics published by the Global Mobile Suppliers Association in June 2022; for these figures and regular updates, refer to GSA (2022).

BIBLIOGRAPHY

3GPP (3rd Generation Partnership Project). 2018. "Technical Report TR 22.822: Study on Using Satellite Access in 5G." 3GPP, Sophia Antipolis Technology Park, France. https://portal.3gpp.org/desktopmodules/Specifications/SpecificationDetails .aspx?specificationId=3372.

5G ACIA. 2019. "5G Non-Public Networks for Industrial Scenarios." *5G Alliance for Connected Industries and Automation*, White Paper, July. https://www.5g-acia.org /fileadmin/5G-ACIA/Publikationen/5G-ACIA_White_Paper_5G_for_Non-Public _Networks_for_Industrial_Scenarios/WP_5G_NPN_2019_01.pdf.

5G Americas. 2020. "5G's Year One: Fast Start and Healthy Growth." Press Release, March 23. https://www.5gamericas.org/5gs-year-one-fast-start-and-healthy-growth/.

Anttonen, Antti, Pekka Ruuska, and Markku Kiviranta. 2019. *3GPP Nonterrestrial Networks: A Concise Review and Look Ahead.* Research Report No. VTT-R-00079-19, VTT Technical Research Centre of Finland. https://cris.vtt.fi/ws/portalfiles/portal /22778833/VTT_R_00079_19.pdf.

Baig, Ghufran, Dan Alistarh, Thomas Karagiannis, Bozidar Radunovic, Matthew Balkwill, and Lili Qiu. 2017. *Towards Unlicensed Cellular Networks in TV White Spaces.* New York: Association for Computing Machinery. doi: 10.1145/3143361.3143367.

Barnes, Robert. 2020. "How to Overcome High Latency in the 5G Era?" *EENews Europe*, May 11. https://www.eenewseurope.com/news/how-overcome-high-latency-5g-era.

Barton, James, ed. 2019. "5G in Emerging Markets." *Developing Telecoms*, October. https://www.developingtelecoms.com/images/reports/5g-em-report-final-1023-4.pdf.

Behar, Rose. 2019. "mmWave vs. Sub-6: The Different Types of 5G and How They Work." *Digi International,* January 6. https://www.digitaltrends.com/mobile/5g-spectrum-variants/.

BEREC (Body of European Regulators for Electronic Communications), and the EU's Radio Spectrum Policy Group (RSPG). 2011. "BEREC-RSPG Report on Infrastructure and Spectrum Sharing in Mobile/Wireless Networks." Document number BoR (11) 26, June 13. https://berec.europa.eu/eng/document_register/subject_matter/berec/download/0/224-berec-rspg-report-on-infrastructure-and-_0.pdf.

Camargos, Luciana. 2019. "The Future of Mobile Is Powered by Great Backhaul." GSM Association. April 16. https://www.gsma.com/spectrum/great-backhaul/#.

Canalys. 2020. "Global Smartphone Shipments Q1 2020." *Newsroom*, April 30. https://www.canalys.com/newsroom/canalys-worldwide-smartphone-shipments-fall-due-to-coronavirus.

Chataut, Robin, and Robert Akl. 2020. "Massive MIMO Systems for 5G and beyond Networks—Overview, Recent Trends, Challenges, and Future Research Direction." *Sensors (Basel)* 20 (10): 2753. https://www.ncbi.nlm.nih.gov/pmc/articles/PMC7284607/.

Cisco. 2018. "Cisco Spearheads Multi-Vendor Open vRAN Ecosystem Initiative for Mobile Networks." Press Release, February 25. https://newsroom.cisco.com/press-release-content?type=webcontent&articleId=1913034.

Clark, Robert. 2019. "Operators Starting to Face Up to 5G Power Cost." *Light Reading*, October 30. https://www.lightreading.com/asia-pacific/operators-starting-to-face-up-to-5g-power-cost-/d/d-id/755255#:~:text=The%20power%20consumption%20of%205G,a%204G%20basestation%2C%20it%20says.&text=Huawei%20estimates%20that%20by%202026,million%20small%20cells%20u.

Communications Update. 2019. "Smart Axiata Trials 5G in Phnom Penh." *Telegeography*, July 9. https://www.commsupdate.com/articles/2019/07/09/smart-axiata-trials-5g-in-phnom-penh/.

Communications Update. 2020a. "Rogers Users Gain 5G Access on Samsung Galaxy S20 Phones." *Telegeography*, March 9. https://www.commsupdate.com/articles/2020/03/09/rogers-users-gain-5g-access-on-samsung-galaxy-s20-phones/.

Communications Update. 2020b. "Telmex Transfers 3.5 GHz Spectrum to Telcel Ahead of 5G Push." *Telegeography*, April 14. https://www.commsupdate.com/articles/2020/04/14/telmex-transfers-3-5ghz-spectrum-to-telcel-ahead-of-5g-push/.

Communications Update. 2020c. [India] "Finance Ministry Critical of 5G Spectrum Pricing." *Telegeography*, May 4. https://www.commsupdate.com/articles/2020/05/04/finance-ministry-critical-of-5g-spectrum-pricing/.

Communications Update. 2020d. [Malaysia] "MCMC Publishes Final Report on Spectrum Allocation Plans; Expects Commercial 5G Rollouts by 3Q20." https://www.commsupdate.com/articles/2020/01/02/mcmc-publishes-final-report-on-spectrum-allocation-plans-expects-commercial-5g-rollouts-by-3q20/.

Communications Update. 2020e. "Russian Ministry Drafts Plan for 5G Network Sharing." *Telegeography*, April 7. https://www.commsupdate.com/articles/2020/04/07/russian-ministry-drafts-plan-for-5g-network-sharing/.

Communications Update. 2020f. [Russia] "24GHz mmWave 5G Spectrum Available without Auction; MegaFon/Rostelecom Receive Test mmWave Band." *Telegeography*, March 20. https://www.commsupdate.com/articles/2020/03/20/24ghz-mmwave-5g-spectrum-available-without-auction-megafonrostelecom-receive-test-mmwave-band/.

Communications Update. 2020g. [Bhutan] "BICMA Publishes 5G Deployment Framework, Aiming for 2022 Start Date." *Telegeography*, March 4. https://www.commsupdate.com/articles/2020/03/04/bicma-publishes-5g-deployment-framework-aiming-for-2022-start-date/.

Communications Update. 2020h. "Vodacom Launches 5G Services in Johannesburg, Pretoria and Cape Town." *Telegeography*, May 5. https://www.commsupdate.com/articles/2020/05/05/vodacom-launches-5g-services-in-johannesburg-pretoria-and-cape-town/.

Cortès, Victor. 2019. "Uruguay Launches Latin America's First 5G Network, El Salvador Trails Behind." *Contxto*, April 11. https://www.contxto.com/en/news/uruguay-launches-latin-americas-first-5g-network-el-salvador-trails-behind/.

DARPA (Defense Advanced Research Projects Agency). 2020. "Open, Programmable, Secure 5G (OPS-5G)." https://www.darpa.mil/program/open-programmable-secure-5g.

DeTomasi, Sheri. 2018. "Understanding 5G New Radio Bandwidth Parts." *Industry Insights Blog*, Keysight Technologies, November 1, 2018. https://blogs.keysight.com/blogs/inds.entry.html/2018/10/31/understanding_5gnew-iYIV.html.

Developing Telecoms. 2018. "Uruguay, Costa Rica and Argentina Lead the Latin American Telecoms Market." July. https://www.developingtelecoms.com/telecom-business/telecom-trends-forecasts/7906-uruguay-costa-rica-and-argentina-lead-the-latin-american-telecoms-market.html.

Developing Telecoms. 2020. "Saudi Announces 5G Smart Campus; UAE Plans for 26 GHz 5G." February. https://www.developingtelecoms.com/telecom-technology/wireless-networks/9205-saudi-announces-5g-smart-campus-uae-plans-for-26ghz-5g.html.

DHS (Department of Homeland Security Science and Technology Directorate). 2021. *Mapping of 5G Technologies and Use Cases to DHS S&T Customer Components*. AOS-21-0579. https://www.dhs.gov/sites/default/files/publications/5g_mapping_may2021_0.pdf.

Donkin, Chris. 2018. "Vodacom Claims Africa First with Lesotho 5G Launch." *Mobile World Live*, August 28. https://www.mobileworldlive.com/featured-content/top-three/vodacom-claims-africa-first-with-lesotho-5g-launch/.

Ericsson. 2020. "Ericsson Mobility Report: November 2020." https://www.ericsson.com/4adc87/assets/local/mobility-report/documents/2020/november-2020-ericsson-mobility-report.pdf.

Essiambre, René-Jean, Gerhard Kramer, Peter J. Winzer, Gerard J. Foschini, and Bernhard Goebel. 2010. "Capacity Limits of Optical Fiber Networks." *Journal of Lightwave Technology* 28 (4): 662–701. https://www.osapublishing.org/jlt/abstract.cfm?uri=jlt-28-4-662.

EU 5G Observatory. 2020a. "5G Connected and Automated Mobility (CAM)." https://5gobservatory.eu/5g-trial/5g-connected-and-automated-mobility-cam/.

EU 5G Observatory. 2020b. "Product/Market Developments." http://5gobservatory.eu/5g-private-licences-spectrum-in-europe/.

Fletcher, Bevin. 2020. "AT&T Adds 5G to 28 New Markets, Expands DSS Deployment." *Fierce Wireless*, June 29. https://www.fiercewireless.com/5g/at-t-expands-5g-to-28-new-markets-continues-dss-deployment.

Flynn, Kevin. 2019. "RAN Rel-16 Progress and Rel-17 Potential Work Areas." *3GPP*, July 18. https://www.3gpp.org/news-events/2058-ran-rel-16-progress-and-rel-17-potential-work-areas.

Forge, Simon, Robert Horvitz, Colin Blackman, and Erik Bohlin. 2019. *Light Deployment Regime for Small-Area Wireless Access Points (SAWAPs)*. European Commission, December. https://op.europa.eu/en/publication-detail/-/publication/463e2d3d-1d8f-11ea-95ab-01aa75ed71a1/language-en/format-PDF/source-112125706.

Getto, Luke. 2019. "The Challenges of 5G Network Densification." *Microwave Journal*. https://www.microwavejournal.com/articles/32235-the-challenges-of-5g-network-densification.

Government of Hong Kong. 2018. *SCED and CA Announce Arrangements for Releasing 5G Spectrum in Various Frequency Bands*. https://www.info.gov.hk/gia/general/201812/13/P2018121300460.htm.

Government of Korea. 2019. *5G+ Strategy to Realize Innovative Growth. Ministry of Science and ICT.* https://www.msit.go.kr/cms/english/pl/policies2/__icsFiles/afieldfile/2020/01/20/5G%20plus%20Strategy%20to%20Realize%20Innovative%20Growth.pdf.

GSA (Global Mobile Suppliers Association). 2020a. *LLTE and 5G Market Statistics— December 2020.* Sheffield, UK: GSA. https://gsacom.com/paper/lte-5g-market-statistics-december-2020/.

GSA (Global Mobile Suppliers Association). 2020b. *LTE and 5G Market Statistics: Global Snapshot April 2020.* Sheffield, UK: GSA.

GSA (Global Mobile Suppliers Association). 2022. *5G Market Snapshot June 2022.* Sheffield, UK: GSA.

GSM Association. 2018. *5G Spectrum: GSMA Public Policy Position.* November. London: GSM Association. https://www.gsma.com/spectrum/wp-content/uploads/2018/12/5G-Spectrum-Positions-1.pdf.

GSM Association. 2019a. *Mobile Backhaul: An Overview.* https://www.gsma.com/futurenetworks/wiki/mobile-backhaul-an-overview/.

GSM Association. 2019b. "Region in Focus: Latin America, Q3 2019." *GSMA Intelligence,* December. https://data.gsmaintelligence.com/research/research/research-2019/region-in-focus-latin-america-q3-2019.

GSM Association. 2019c. "5G in Sub-Saharan Africa: Laying the Foundations." *GSMA Intelligence.* https://www.gsma.com/subsaharanafrica/resources/5g-in-sub-saharan-africa-laying-the-foundations.

GSM Association. 2019d. *The 5G Guide.* https://www.gsma.com/wp-content/uploads/2019/04/The-5G-Guide_GSMA_2019_04_29_compressed.pdf.

GSM Association. 2020a. "Global 5G Landscape, Q1 2020." *GSMA Intelligence,* April. https://data.gsmaintelligence.com/research/research/research-2020/global-5g-landscape-q1-2020.

GSM Association. 2020b. "Global Forecast Review: Q1 2020 Updates." *GSMA Intelligence,* April. https://data.gsmaintelligence.com/research/research/research-2020/global-forecast-review-q1-2020-updates.

GSM Association. 2020c. *The Mobile Economy 2020.* https://www.gsma.com/mobileeconomy/wp-content/uploads/2020/03/GSMA_MobileEconomy2020_Global.pdf.

GSM Association. 2020d. *The Mobile Economy: Middle East and North Africa 2019.* June. https://www.gsma.com/mobileeconomy/wp-content/uploads/2020/03/GSMA_MobileEconomy2020_MENA_Eng.pdf.

GSM Association. 2020e. "Region in Focus: North America. Q4 2019." *GSMA Intelligence,* January. https://data.gsmaintelligence.com/research/research/research-2020/region-in-focus-north-america-q4-2019.

GSM Association. 2020f. "Global Mobile Trends 2021: Navigating Covid-19 and Beyond." *GSMA Intelligence,* December. https://data.gsmaintelligence.com/api-web/v2/research-file-download?id=58621970&file=141220-Global-Mobile-Trends.pdf.

GSM Association. 2020g. "GSMA Intelligence Operator Device Survey." https://data.gsmaintelligence.com/api-web/v2/research-file-download?id=51249714&file=130520-Device-Portfolios-2.pdf.

GSM Association. 2021a. "Global 5G Landscape, Q4 2020." *GSMA Intelligence,* January. https://data.gsmaintelligence.com/research/research/research-2021/global-5g-landscape-q4-2020.

GSM Association. 2021b. "5G Spectrum: GSMA Public Policy Position." March. GSM Association, London. https://www.gsma.com/spectrum/wp-content/uploads/2021/04/5G-Spectrum-Positions.pdf.

GSM Association. 2021c. "Wireless Backhaul Evolution: Delivering Next-Generation Connectivity." February. GSM Association, London. https://www.gsma.com/spectrum/wp-content/uploads/2022/04/wireless-backhaul-spectrum.pdf.

GSM Association. 2023. "5G Global Launches & Statistics." GSM Association, London. https://www.gsma.com/futurenetworks/ip_services/understanding-5g/5g-innovation/.

Handford, Richard. 2019. "Setback for Indian Spectrum Sale as Government and Regulator Disagree." *PolicyTracker*, July 24. https://www.policytracker.com/setback-for-indian-spectrum-sale-as-dot-trai-disagree/.

Harb, Robbie. 2020. "Korea Announces Tech New Deal to Stoke Post-Coronavirus Economy." *The Register.* https://www.theregister.co.uk/2020/05/11/south_korea_tech_new_deal/.

Hardesty, Linda. 2019. "SBA Says Sprint Is the Only Carrier It Sees Deploying Massive MIMO." *Fierce Wireless*, October 29. https://www.fiercewireless.com/5g/sba-says-sprint-only-carrier-it-sees-deploying-massive-mimo.

Headrick, D. 2019. "5G Modularity." *Compass Magazine,* October 30. https://compassmag.3ds.com/5g-modularity/.

Hilbert, Martin. 2016. "The Bad News Is That the Digital Access Divide Is Here to Stay: Domestically Installed Bandwidths among 172 Countries for 1986–2014." *Telecommunications Policy.* http://doi.org/10.1016/j.telpol.2016.01.006.

Horwitz, Jeremy. 2018. "World's 'First' Commercial 5G Network Launches in Qatar." *Venture Beat,* May 14. https://venturebeat.com/2018/05/14/worlds-first-commercial-5g-network-launches-in-qatar/.

Huawei. 2019. "World's First Remote Operation Using 5G Surgery." Press Release, January 17. https://www.huawei.com/us/industry-insights/outlook/mobile-broadband/wireless-for-sustainability/cases/worlds-first-remote-operation-using-5g-surgery.

Huawei. 2020. "5G Top 10 Use Cases." https://www.huawei.com/us/industry-insights/outlook/mobile-broadband/xlabs/use-cases/5g-top-10-use-case.

Hwang, Chang-Gyu. 2019. *5G Commercialization and the Future Economic Impact.* Commissioner Insights, Broadband Commission. http://reports.broadbandcommission.org/state-of-broadband-2019/chapter-6/overlay/insight-from-commissioner-dr-chang-gyu-hwang-korea-telecom/.

IHS Markit. 2019. "The 5G Economy: How 5G Will Contribute to the Global Economy." https://www.qualcomm.com/content/dam/qcomm-martech/dm-assets/documents/the_ihs_5g_economy_-_2019.pdf.

ITU (International Telecommunication Union). 2007–2012–2015. *Resolution ITU-R 56-2: Naming for International Mobile Telecommunications.* https://www.itu.int/dms_pub/itu-r/opb/res/R-RES-R.56-2-2015-PDF-E.pdf.

ITU (International Telecommunication Union). 2015. *Recommendation ITU-R M.2083-0: IMT Vision—Framework and Overall Objectives of the Future Development of IMT for 2020 and Beyond.* Geneva: ITU. https://www.itu.int/dms_pubrec/itu-r/rec/m/R-REC-M.2083-0-201509-I!!PDF-E.pdf.

ITU (International Telecommunication Union). 2017. *Report ITU-R M.2412-0: Guidelines for Evaluation of Radio Interface Technologies for IMT-2020.* https://www.itu.int/dms_pub/itu-r/opb/rep/R-REP-M.2412-2017-PDF-E.pdf.

Johns Hopkins Applied Physics Laboratory. 2021. "Mapping of 5G Technologies and Use Cases to DHS S&T Customer Components." Report AOS-21-0579, Johns Hopkins Applied Physics Laboratory, Laurel, MD. https://www.dhs.gov/sites/default/files/publications/5g_mapping_may2021_0.pdf.

Kajastie, Nia. 2019. "Sweden's First 5G Network for Industry Goes Live." *Mining Magazine*, March 12. https://www.miningmagazine.com/communications/news /1358410/swedens-first-5g-network-for-industry-goes-live.

Kelly, Chris. 2020. "Etisalat to Trial OpenRAN Solutions across MEA." *CommsMEA*, February 20. https://www.commsmea.com/technology/infrastructure/21437 -etisalat-to-trial-openran-solutions-across-mea.

Klaehne, Maurice. 2020. "5G Smartphones Account for 14% of Total US Smartphone Sales in Aug." *Counterpoint Research*, Press Release, October 7. https://www.counterpointresearch.com/5g-smartphones-account-for-14-of-total -us-smartphone-sales-in-aug/.

Kurbatov, Dmitry. 2020. "Private 5G Networks Aren't Immune to Cyber Attacks." *5G Radar*, August 11. https://www.5gradar.com/features/private-5g-networks -arent-immune-to-cyber-attacks.

Lee, Jonn Jun. 2020. "Huawei Widens Telecom Share Gap with Nokia." *The ELEC: Korea Electronics Industry Media*, September 9. http://www.thelec.net/news /articleView.html?idxno=1522.

Lennighan, Mary. 2019. "Maldives Gets 5G, But It's Not for the Locals." *TelecomTV*, August 6. https://www.telecomtv.com/content/5g/maldives-gets-5g-but-its-not-for -the-locals-35949/.

Lombardi, Renato. 2019. "Wireless Backhaul for IMT-2020/5G: Overview and Introduction." Presented at ITU Workshop on the Evolution of Fixed Service in Backhaul Support of IMT 2020/5G, Geneva, April 29. https://www.itu.int/en/ITU-R /study-groups/workshops/fsimt2020/Documents/1-Wireless%20Backhaul%20 for%20IMT%202020-5G%20-%20Overview%20and%20introduction.pdf.

Lu, Lu, Geoffrey Li, A. Lee Swindlehurst, and Alexei Ashikhmin. 2014. "An Overview of Massive MIMO: Benefits and Challenges." *IEEE Journal of Selected Topics in Signal Processing* 8 (5): 742.

Mavenir. 2019. *Virtualisation Paves the Way for Affordable 5G*. https://www.raconteur .net/sponsored/virtualisation-paves-the-way-for-affordable-5g.

Mavenir. 2020. "OpenRAN Is an Alternative Way of Building Networks that Promises Greater Interoperability and More Competition." https://mavenir.com/portfolio /access-edge-solutions/radio-access/openran/.

McKinsey & Company. 2018. *Network Sharing and 5G: A Turning Point for Lone Riders*. https://www.mckinsey.com/industries/technology-media-and-telecommunications /our-insights/network-sharing-and-5g-a-turning-point-for-lone-riders.

McKinsey Global Institute. 2020. *Connected World: An Evolution in Connectivity Beyond the 5G Revolution*. https://www.mckinsey.com/industries/technology-media -and-telecommunications/our-insights/connected-world-an-evolution -in-connectivity-beyond-the-5g-revolution.

Mitsubishi Electronics. 2016. "Mitsubishi Electric's New Multibeam Multiplexing 5G Technology Achieves 20Gbps Throughput." Press Release No. 2984, January 21. https://emea.mitsubishielectric.com/en/news-events/releases/global/2016/0121 -a/index.html.

Morris, Ian. 2018. "Major Telcos Pool Efforts to Slash 5G RAN Costs." *Light Reading*. February 27. https://www.lightreading.com/mobile/fronthaul-c-ran/major-telcos -pool-efforts-to-slash-5g-ran-costs/d/d-id/740913.

Moura Gomes, Andre. 2021. *Global Trends: 5G* (Report No: CTGTGO20210005). Brussels: Cullen International. https://www.cullen-international.com/client /site/documents/CTGTGO20210005?version=this.

Myers, Margaret, and Guillermo Garcia Montenegro. 2019. "Latin America and 5G: Five Things to Know." *The Dialogue* (Inter-American Dialogue). https://www .thedialogue.org/analysis/latin-america-and-5g-five-things-to-know/.

Naar, Ismaeel. 2020. "Saudi Arabia Ranks Third Globally in Deployment of 5G Technology: Ministry." *Al Arabiya*, February 23. https://english.alarabiya.net /en/business/technology/2020/02/23/Saudi-Arabia-ranks-third-globally -in-deployment-of-5G-technology-Ministry.

Newson, Paul, Hemang Parekh, and Harpinder Matharu. 2018. "Realizing 5G New Radio Massive MIMO Systems." *EDN Asia*. January 18. https://www.ednasia.com /realizing-5g-new-radio-massive-mimo-systems/.

Ngila, Faustine. 2020. "Tanzania, Kenya in Line for Fibre Optic Hook-Up." *The East African*, May 16. https://www.theeastafrican.co.ke/news/ea/Tanzania-Kenya-in -line-for-fibre-optic-hook-up/4552908-5554606-gajvta/index.html.

NICT (National Institute of Information and Communications Technology). 2020. "First Transmission of 1 Petabit/S Using a Single-Core Multimode Optical Fiber." *SciTech Daily*, December 18. https://scitechdaily.com/worlds-first-transmission-of -1-petabit-s-using-a-single-core-multimode-optical-fiber/.

Nordrum, Amy. 2016. "Millimeter Waves Travel More Than 10 Kilometers in Rural Virginia 5G Experiment." *IEEE Spectrum*, November 7. https:// spectrum.ieee.org/tech-talk/telecom/wireless/millimeter-waves-travel-more -than-10-kilometers-in-rural-virginia.

nPerf. 2020. *5G Coverage Map Worldwide* (generated hourly). https://www.nperf.com /en/map/5g.

Office of the Communications Authority, Hong Kong. 2020. *Subsidy Scheme for Encouraging Early Deployment of 5G.* https://www.ofca.gov.hk/en/industry_focus /telecommunications/5g_subsidy/index.html.

ONF (Open Network Foundation). 2020. *2020 State of the ONF.* https://www .opennetworking.org/news-and-events/blog/2020-state-of-the-onf/.

Oughton, Ed, Vivien Foster, Jim W. Hall, and Niccolo Comini. 2022. "Policy Choices Can Help Keep 4G and 5G Universal Broadband Affordable." *Technological Forecasting and Social Change* 176: 121409. doi:10.1016/j.techfore.2021.121409.

Ovum. 2018. *Fixed Wireless Access: Changing the Face.* https://www.omdia.com/~/media /informa-shop-window/tmt/whitepapers-and-pr/fixed-wireless-access.pdf.

Oxford Economics. 2019. *Restricting Competition in 5G Network Equipment: An Economic Impact Study.* December. https://resources.oxfordeconomics.com/hubfs /Huawei_5G_2019_report_V10.pdf.

Peisa, Janne, Patrik Persson, Stefan Parkvall, Erik Dahlman, Asbjørn Grøvlen, Christian Hoymann, and Dirk Gerstenberger. 2020. "5G Evolution: 3GPP Releases 16 & 17 Overview." *Ericsson Technology Review*, March 9. https://www.ericsson.com/en /reports-and-papers/ericsson-technology-review/articles/5g-nr-evolution.

Perez, Bien. 2017. "Why China Is Set to Spend US$411 Billion on 5G Mobile Networks." *South China Morning Post*, 19 June. https://www.scmp.com/tech/china-tech/article /2098948/china-plans-28-trillion-yuan-capital-expenditure-create-worlds.

Polaris Market Research. 2020. "5G Smartphone Market." *Market Research Report*, PM1707. https://www.polarismarketresearch.com/industry-analysis/5g-smartphone -market.

PwC. 2021. "The Global Economic Impact of 5G." *PwC*. https://www.pwc.com/gx/en /tmt/5g/global-economic-impact-5g.pdf.

Remmert, Harald. 2022. "LTE vs. 5G: What Is the Difference?" *Digi International*, May 4. https://www.digi.com/blog/post/lte-vs-5g.

Rogers Communications. 2020. "Rogers Only Carrier to Offer 5G Service on Canada's First 5G Smartphone." Press Release, March 6. https://about.rogers.com/news-ideas /rogers-only-carrier-to-offer-5g-service-on-canadas-first-5g-smartphone/.

Rosenberg, Leslie. 2021. "Private Enterprise 5G/LTE Mobile Networks Are Team Sports & Services Firms Are the Captains." *IDC Blog*, March 21. https://blogs .idc.com/2021/03/03/private-enterprise-5g-lte-mobile-networks-are-team -sports-services-firms-are-the-captains/.

Scrase, Adrian. 2018. "3GPP Overview: The Standardization Ecosystem for Global Mobile Systems." Presented at the 3GPP Summit, Makuhari Messe, October 17. https://www.3gpp.org/ftp/Information/presentations/presentations_2018/2018_10_17_tokyo/presentations/2018_1017_3GPP%20Summit_02_Key%20Note_SCRASE.pdf.

Si, Ma. 2020. "Realme Launches 5G Phone for $200." *China Daily*, August 4. https://www.chinadaily.com.cn/a/202008/04/WS5f28b8b6a31083481725e01a.html.

Siddika, Fatema, Md Anwar Hossen, and Sajeeb Saha. 2017. "Transition from IPv4 to IPv6 in Bangladesh: The Competent and Enhanced Way to Follow." 2017 International Conference on Networking, Systems and Security (NSysS), Dhaka, Bangladesh, 2017, pp. 174–9. IEEE.

Smee, John E. 2016. "5 Wireless Inventions that Are Making 5G NR—The Global 5G Standard—A Reality." *Qualcomm's OnQ Blog*, December 20. https://www.qualcomm.com/news/onq/2016/12/20/5-wireless-inventions-are-making-5g-nr-global-5g-standard-reality.

Song, Steve. 2016. "What If Everyone Had Free 2G Mobile Internet Access?" *ICTWorks*, January 6. https://www.ictworks.org/what-if-everyone-had-free-2g-mobile-internet-access/.

Strategy Analytics. 2020. *Strategy Analytics: Samsung Leads in 5G Smartphones in Q1 2020.* April 28. https://news.strategyanalytics.com/press-releases/press-release-details/2020/Strategy-Analytics-Samsung-Leads-in-5G-Smartphones-in-Q1-2020/default.aspx.

Stryjak, Jan. 2020. "Region in Focus: Asia Pacific, Q4 2019." *GSMA Intelligence*, March. https://data.gsmaintelligence.com/research/research/research-2020/region-in-focus-asia-pacific-q4-2019.

Taleb, Tarik, Konstantinos Samdanis, Badr Mada, Hannu Flinck, Sunny Dutta, and Dario Sabella. 2017. "On Multi-Access Edge Computing: A Survey of the Emerging 5G Network Edge Cloud Architecture and Orchestration." *IEEE Communications Surveys & Tutorials* 19 (3). http://www.mosaic-lab.org/uploads/papers/c191e2bf-70d4-40ed-ba6d-e82f0c4c156c.pdf.

Tang, Frank. 2020. "Coronavirus: China, Xi Jinping Put 5G Technology on Top of Huge Spending Plans to Salvage Economy." *South China Morning Post.* https://www.scmp.com/economy/china-economy/article/3065229/coronavirus-china-xi-jinping-put-5g-technology-top-huge.

Tikhvinskiy, Valery, and Victor Koval. 2020. "Prospects of 5G Satellite Networks Development." In *Moving Broadband Mobile Communications Forward: Intelligent Technologies for 5G and Beyond*, edited by Abdelfatteh Haidine, 83–98. London: IntechOpen. doi:10.5772/intechopen.90943.

Tomás, Juan Pedro. 2019. "Germany Opens Process for Private 5G Licenses." *RCR Wireless*, November 21. https://www.rcrwireless.com/20191121/5g/germany-opens-process-for-private-5g-licenses.

United Nations. 2020. *Sustainable Development Goals.* https://sustainabledevelopment.un.org/?menu=1300.

University of Oulu. 2020. *6G Waves Magazine*, issue 1 (spring). http://jultika.oulu.fi/files/isbn9789526225838.pdf.

Van Dijk, J.A.G.M. 2006. "Digital Divide Research, Achievements and Shortcomings." *Poetics* 34 (4-5): 221–35. doi:10.1016/j.poetic.2006.05.004.

Vanshika. 2020. *India's 5G Technology (TSDSI-RIT) Moves another Step Forward at ITU.* Telecommunications Standards Development Society of India. https://tsdsi.in/indias-5g-technology-tsdsi-rit-moves-another-step-forward-at-itu/.

Viasat. 2019. "Satellite Systems and the 5G Ecosystem." Presented at the ITU International Satellite Symposium, October. https://www.itu.int/en/ITU-R/space/workshops/2019-SatSymp/Presentations/101%20%205G%20Satellites%20Viasat-GSC-GVF.pdf.

Waring, Joseph. 2020. "Vietnam Positioned for Early Move to 5G." *Mobile World Live*, May 15. https://www.mobileworldlive.com/asia/asia-news/vietnam-positioned -for-early-move-to-5g/.

Weissberger, Alan. 2021. "Massive MIMO: Mavenir and Xilinx Collaborate on Massive MIMO for Open RAN." *IEEE Communications Society Technology Blog*, April 13. https://techblog.comsoc.org/tag/massive-mimo/.

Wilson, Stephen. 2020. "Early 5G Fixed-Wireless Access Retail Offers Have Yet to Truly Disrupt the Fixed Broadband Market." *Analysys Mason*, January 21. https:// www.analysysmason.com/research/content/comments/5g-fwa-retail-rdmb0/.

Wood, Ruper. 2019. "5G Fixed–Wireless Access." Analysys Mason, March. https:// www.analysysmason.com/globalassets/x_migrated-media/media/analysys _mason_5g_fixed_wireless_mar2019_samples_rdns03.pdf.

Yifei, Yuan, Zhao Yajun, Zong Baiqing, and Parolari Sergio. 2020. "Potential Key Technologies for 6G Mobile Communications." *Science China Information Sciences* 63: 183301. https://doi.org/10.1007/s11432-019-2789-y.

Yonhap News Agency. 2020. *Monthly Increase in 5G Subscribers Reaches Record High in May: Data*. June 30. https://en.yna.co.kr/view/AEN20200630009500320.

2

5G-Enabled Economic Growth and Potential Applications for Smart Cities, Energy, Transport, and Agriculture

KEY MESSAGES

- Fifth-generation (5G) mobile network technologies have enhanced functionality that potentially can affect industry and the economy in powerful ways. On a global scale, manufacturing, government, health care, retail, and information technology are expected to experience the greatest positive effects from 5G integration.

- 5G supports many use cases, which have varying requirements for speed and latency.

  - In the *urban sector*, 5G can accelerate the emergence of smart cities through the sophisticated management of integrated urban systems and services.

  - In the *energy sector*, 5G can provide enhanced capability to balance supply and demand and accommodate flexible renewable resources.

  - In the *transport sector*, 5G can support the move toward fully autonomous vehicles and intelligent transport systems.

  - In *agriculture*, 5G can support the shift to a "precision" model of real-time, in-field optimization, leading to enhanced productivity, better market coordination, and reduced food loss from spoilage and waste.

- Despite these attributes, 5G is unlikely to be the most critical driver of digital transformation across these sectors or the overall digital economy; rather, it provides an additional layer of advanced mobile connectivity alongside other modes of network connectivity in support of new, innovative applications that can benefit from 5G's ultra-low latency and high data throughput. In circumstances in which alternative modes of connectivity, such as fiber, are scarce or unavailable, 5G could play an important role in facilitating broad digital transformation.

• Features of 5G's technology and performance capabilities, resulting, for example, from network slicing, have generated much excitement. However, new use cases with 5G are difficult to predict. Commercial use cases drive transformational change, as with third-generation (3G) mobile network technologies and subsequent developments with handsets and applications.

• Further research is needed to identify the economic impacts of 5G in developing countries, overall and within key industry verticals, to help decision-makers and other stakeholders understand this technology's role in economic development.

BENEFITS AND POTENTIAL IMPACT OF 5G ADOPTION

5G's enhanced functionality, as compared with that of earlier generations, suggests that it could have a significant impact on industry and broader economic development. Some of these impacts can be considered as extensions and enhancements of current mobile technology. Others, such as integrating 5G into specialized industrial applications, will be new and will likely have a significant impact. Much is still unknown, as applications and innovative uses for 5G are being continuously developed throughout the world.

The deployment and integration of 5G network technology in industry is predicted to have a greater impact on many sectors, particularly manufacturing, as compared with earlier generations. Key sectors that are forecast to benefit the most from 5G include manufacturing and retail, as well as the information and communications, public services, construction, transport, health care, and agriculture sectors (KPMG International 2019). Figure 2.1 presents sales forecasts for 2035, estimated by IHS Markit, across a variety of sectors to be unlocked by the integration of 5G technology. The figure illustrates the magnitude of the anticipated impact on manufacturing as compared with other sectors.

Table 2.1 provides an overview of 5G use cases across six vertical industries where this generation may have the potential to accelerate digital transformation in the form of new business uses and improvements in productivity and efficiency. Many of the verticals in the table do not require 5G to deliver the stated benefits in the right-side column, such as remote health care and mission-critical public safety applications, which can deliver these benefits using earlier mobile network technologies.

FIGURE 2.1 **Industry sales enabled by 5G: Forecasts for 2035**

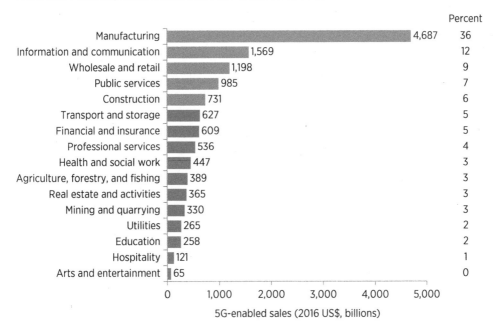

	5G-enabled sales (2016 US$, billions)	Percent
Manufacturing	4,687	36
Information and communication	1,569	12
Wholesale and retail	1,198	9
Public services	985	7
Construction	731	6
Transport and storage	627	5
Financial and insurance	609	5
Professional services	536	4
Health and social work	447	3
Agriculture, forestry, and fishing	389	3
Real estate and activities	365	3
Mining and quarrying	330	3
Utilities	265	2
Education	258	2
Hospitality	121	1
Arts and entertainment	65	0

Source: IHS Markit 2019.

TABLE 2.1 **Potential digital transformation in selected vertical sectors**

Vertical sector	5G use cases facilitating digital transformation	Potential outcomes and benefits
Transportation	5G data from smart vehicles may enable increased public transport efficiency, address complex traffic situations, and decrease congestion.	• Optimized transport routes and improved commutes • Reduced pollution • Decreased fatalities
Health care	5G and better-connected health care devices can provide a platform to develop virtual training, remote cooperation, and remote diagnostic services.	• Improved health care quality and access • Lower health care costs • More wearable sensors and devices
Education	5G, along with virtual reality and augmented reality, provides a platform on which remote access to high-quality learning applications can be developed, especially for activities requiring fine manual skills, such as surgery.	• Improved access and availability • Improved quality (due to real-time feedback during training)
Public safety	5G provides a platform for the development of applications to deliver mission-critical communications, as well as enabling connected ambulances and drones.	• Faster, more efficient emergency responses • Increased safety • Better remote monitoring of developing events
Industry (manufacturing)	5G provides a platform for developing applications for smart factories, improving efficiency and quality assurance, while remote control will minimize danger to operators.	• More efficient production processes • Increased safety
Agriculture	5G provides a platform for the development of applications for smart agriculture to improve the efficiency of processes and sustainable farming through better remote monitoring and automation.	• More efficient agricultural production • Reduced waste • More environmental sensors and devices

Source: Wilson 2020.
Note: 5G = fifth generation mobile network technologies.

5G AND INDUSTRY 4.0

Several countries have recently embraced the notion of the Fourth Industrial Revolution—known as "Industry 4.0"—by developing economic policies aimed at creating an ecosystem where manufacturing systems go beyond connectivity. These systems are now capable of integrating the internet of things (IoT), data analytics, robotics, artificial intelligence, advanced materials, and augmented reality into a single industrial vision.

5G has an edge over fourth-generation (4G) mobile network technologies and Wi-Fi technologies in terms of latency and capacity to handle more connected devices, including high-bandwidth IoT. Through these pathways, 5G may enable the disruptive business models anticipated in Industry 4.0. Although great interest exists in Industry 4.0, doubts remain as to the timing and size of the opportunity.

For a manufacturing-intensive country like China, the Industry 4.0 vision is expressed in the "Made in China 2025" and "Internet Plus" strategies, in which robust capabilities in telecoms, cloud computing, IoT, and big data loom large (Xinhuanet 2019). Other 5G frontrunners in Asia, such as Japan, the Republic of Korea, and Singapore, as well as some European countries and the United States, have also adopted Industry 4.0 concepts.

In 2019, the United Nations Conference on Trade and Development (UNCTAD) explored developing countries' embrace of Industry 4.0, focusing on its direct and indirect impacts on economic and social inequality. UNCTAD notes that "the network nature of digital applications that comprise Industry 4.0 increasingly results in a world in which winners take all, causing high levels of market concentration" (UNCTAD 2019, 6). For example, a few technology companies with access to large amounts of data and funding may succeed in dominating certain technologies and markets. UNCTAD further observes that Industry 4.0 may affect employment in both developed and developing countries and may increase the technological gaps between them.

The pattern and magnitude of the impact of Industry 4.0 on economies and labor markets is still being debated. In cases in which 4G or Wi-Fi 6 connectivity are insufficient to deliver the low latency and high throughput needed, it remains unclear whether a proactive 5G approach or one that awaits further evidence from use cases in industrial settings in early adopter countries will be most appropriate.

5G AND THE SDGs

Beyond Industry 4.0, 5G has the potential to help countries progress toward the Sustainable Development Goals (SDGs). The SDGs were conceived in September 2015 when world leaders attending the United Nations General Assembly adopted the "2030 Agenda on Sustainable Development," which proposed an integrated set of 17 SDGs to be achieved by 2030. Although many SDGs can be facilitated by mobile internet connectivity, additional opportunities may exist to accelerate achievement of the goals and targets through 5G network access and adoption.

The World Economic Forum (2020) discusses 5G's potential contributions to the SDGs, finding that social value can be delivered across 11 key areas, including SDG Goal 3 (good health and well-being) and SDG Goal 9 (industry, innovation, and infrastructure). Other goals to which 5G connectivity is expected to contribute include SDG Goal 8 (decent work and economic growth), SDG Goal 11 (sustainable cities and communities), and SDG Goal 12 (responsible consumption and production).

However, detailed assessments of 5G's impacts are few as of 2023 given the ongoing state of development and deployment and, therefore, can only be speculated on qualitatively through pilot tests and use cases. With this issue in mind, this chapter explores how 5G could have a role in four sectors linked to key SDGs, specifically the following:

- Safer, more resilient, and sustainable smart cities (SDG Goal 11 on sustainable cities and communities);

- More affordable, reliable, sustainable, and modern energy (SDG Goal 7 on affordable and clean energy);

- Transport sector improvements that support many SDGs; and

- Sustainable agriculture (SDG Goal 2 on zero hunger).

5G USE CASES IN FOUR INDUSTRY VERTICALS

As of 2023, the use cases enabled by 5G are still being developed and explored through sectoral trials and pilots, most of which are in highly developed countries that were early adopters of 5G technologies. Exploring them in more detail can provide a a sense of what may be possible when deploying 5G applications in low- and middle-income countries.

Most of the ongoing 5G trials and pilots are driven by partnerships formed among industrial firms, mobile network operators, and equipment vendors operating globally. Prominent examples are Ericsson, Huawei, Nokia, and Samsung Electronics. Operators and equipment vendors contribute their expertise in network supply and operation, while industrial firms contribute their domain knowledge. Large corporations, such as Bosch and Siemens, have chosen to build and manage their own 5G networks, taking advantage of 5G's capabilities to set up local private networks. Smaller enterprises are more likely to partner with mobile network operators or system integrators until the 5G ecosystem becomes sufficiently mature for them to consider deploying private 5G networks as a cost-effective solution (Lee, Casey, and Wigginton 2019).

Several characteristics of 5G suggest that some enterprises and industries could benefit even before the networks are fully deployed for public use. The sectors likely to benefit include urbanization and smart cities, energy, transport, and agriculture.

Urbanization and Smart Cities

Smart cities use data and digital technologies to plan and manage their core functions and become more efficient, innovative, inclusive, and resilient (World Bank 2015). The process of creating smoother and safer living conditions in cities through a gradual digital transformation has already begun, particularly in China and Korea. It has also begun on a much smaller scale in large cities across the Global South in verticals such as smart lighting or smart parking.

Improving urban environments is an urgent task facing developing countries, and smart infrastructure can be key to accelerating and revolutionizing urban development. Many cities in developing countries are equipped with smart infrastructure. For example, in Africa, these cities include. Libreville in Gabon; Nairobi in Kenya; Abuja, Calabar, and Lagos in Nigeria; Cape Town and Johannesburg in South Africa; and Kampala in Uganda. In South America, notable examples are Santa Cruz de la Sierra in Bolivia, Guayaquil in Ecuador, and Claro and Trujillo in Peru.

Although 5G is not needed to connect IoT devices, its high capacity and low latency, which were discussed in chapter 1, can speed certain smart city processes and features—particularly where alternatives such as fiber are limited or unavailable. For example, while broader mobile ecosystems can be used to allow information gathered through sensors to be transmitted in real time to central monitoring locations, the standards set for 5G imply that it could have the capacity for far more endpoint devices to communicate in real time. This innovation would be possible due to massive machine-type communications technology and related capabilities known as "critical machine-type communication."

Massive machine-type communications are intended for large numbers of IoT devices, which in smart cities are sensors and actuators that transmit data back and forth. Sample applications include smart buildings (smart energy systems), logistics and fleet management, and air and water quality monitoring. Massive machine-type communications are designed to be latency tolerant, efficient for transmitting and receiving small data blocks, and suitable for transmission over low-bandwidth pipes (Beheshti 2019). Critical machine-type communication is intended for applications that are mission critical and require accurate data transmission in real time. Examples include traffic safety control and management of the electrical grid. Smart city solutions applied to the management of vehicle traffic and electrical grids alone could produce considerable benefits and savings through reductions in energy use, traffic congestion, and fuel costs.

Mobile broadband will be a key foundational layer for deploying smart infrastructure, particularly where fiber broadband infrastructure is insufficient. These mobile-based strategies are certain to involve multiple generations of network technology, such as 4G long-term evolution, but 5G can provide additional opportunities where ultra-low latency and increased network capacity are needed. Many smart city applications require low-power sensors to send relatively small amounts of information, such as temperature or water levels, which can be connected through other technologies such as NarrowBand–internet of things (NB-IoT), wireless smart ubiquitous networks or, in some cases, even second-generation (2G) mobile network technologies.

Although 5G is not required to achieve smart city objectives given that other technologies are available, a 5G-enabled smart city will be less constrained in the types and quantities of devices that can be connected. Developing countries that aim to make progress toward SDG Goal 11 (sustainable cities and communities) should explore how 5G connectivity might help them meet their goals but also ensure that other available technologies are considered, to compare and assess both the value proposition and viability.

Energy

As with smart cities and urbanization, the energy sector is undergoing a major digital transformation that can benefit from the advanced capabilities of 5G connectivity. The electric power industry traditionally has been at the forefront of digital innovation and infrastructure in energy, as the complex exercise of balancing supply and demand in real time has led to the use of advanced equipment and software for planning and operational control (for example, smart grids 1.0, supervisory control and data acquisition, digital image correlation, and atomic force microscopy). Many utilities have installed their own information and communication technology and have even become mobile network operators.

Currently, this industry relies on a combination of wired and wireless technologies for a range of applications that are needed to deliver electricity services—power-line communication, NB-IoT, fiber or digital subscriber line, wireless radio, Wi-Fi, cellular (3G and 4G), and satellite. This technology is adapted to the task of delivering electric power to consumers in systems in which the consumers largely are passive. On the consumer side, the most common application is related to smart metering, which already has been provided using older mobile generations as well as NB-IoT. On the operational side, applications such as managing substations are generally achieved through lower-performance connectivity technologies given the low latency requirements.

The latest digital transformation in energy is characterized by a growing share of variable renewable energy (such as solar and wind power), the emergence of distributed energy resources operated by "prosumers" (people who both produce and consume electricity), and significant changes in the composition and complexity of the demand load (for example, through the increased use of cooling technologies and charging of electric vehicles). New technologies and practices are shifting the customary, centralized industry built around a one-way flow of electricity into a two-way flow system—or two-sided market—with multiple players producing, storing, and trading energy services (Fan and Son 2020).

The new operating environment and shifting demands require an over-haul of traditional processes. Ways must be found to handle the planning, monitoring, operation, and control of millions or even billions of energy devices in real time, most of which will involve the collection, storage, and processing of big data. In this setting, 5G and complementary Industry 4.0 digital technologies can support the transformation of the energy industry, but 5G is unlikely to be a revolutionary driver. 5G would be useful for trans-mitting large data volumes where needed to perform advanced and predictive analytics based on complex processing algorithms and artificial intelligence, which can be used for better predicting and operating a multilayered electric-ity system that relies on big data processing. Table 2.2 describes three electric-ity use cases highlighting how 5G connectivity can amplify cost savings and other benefits.

The potential benefits that 5G infrastructure could bring to the energy sector fall into two broad categories: (1) enhancing operational and economic efficiency of utility services and (2) amplifying the value of decentralized energy resources. In the first category, 5G connectivity can support the management of distributed energy resources, a source of decentralized, community-generated energy. Distributed energy resources include behind-the-meter renewable and nonrenewable generation,

TABLE 2.2 Three examples of the beneficial use of 5G in the power industry

Use case	Description
Remote monitoring and maintenance of facilities, infrastructure, and personnel	By equipping their plants with networked sensors, energy companies can continuously collect and analyze data on the condition and operation of facilities, sometimes enabling the companies to detect when maintenance is required before a piece of equipment fails, thus minimizing downtime. In some cases, maintenance can be performed remotely. Personnel can be monitored with internet of things–connected biometry to limit their exposure to hazardous situations.
Advanced (smart) consumption metering	Energy companies are already replacing traditional mechanical meters with "smart meters." Connected to the internet, smart meters can transmit, process, and receive a range of operational data that are useful for both energy suppliers and customers. Benefits include an end to the need for energy companies to dispatch meter readers and the empowerment of customers seeking to make informed decisions about their energy use and suppliers. Many smart meters will be adopted worldwide in the next few years. Now that the NarrowBand–internet of things is available, the technology of choice for communicating with these smart meters is international mobile telecommunications.
Grid asset management and protection	Energy generation and distribution grids are expensive to build and maintain. Energy companies strive to reduce the cost of grid construction and operation while maximizing the return on their investment. In addition, they must satisfy stricter requirements stemming from distributed generation, tighter regulation, and more sophisticated security threats. By equipping the grid with connected sensors and actuators, operators can automate a wider range of maintenance and security responses. They can also collect, analyze, and act on data from the grid to optimize performance continuously. As well as being used by the energy companies themselves, such data can be shared as appropriate with business partners and customers.

Sources: 5G-PPP 2018a, 2018b, 2018c.

energy storage, inverters (devices that convert direct current into alternating current), electric vehicles, and other controlled loads (for example, separately metered appliances like hot water systems). Distributed energy resources also use new technologies such as smart meters and data services. Common examples include rooftop photovoltaic solar units, natural gas turbines, microturbines, wind turbines, biomass generators, fuel cells, battery storage, electric vehicles and their chargers, and demand-response applications. Together, these elements constitute distributed generation (ARENA 2018).

Most existing plans for managing distributed energy resources are based on existing network connectivity technology, including earlier mobile network generations and fiber. 5G's main role would be to add an additional layer of high-capacity connectivity to support new applications rather than replacing alternative network connections where they exist.

5G has the potential to contribute to the efficient management of distributed energy resources by enabling real-time system communications. Decentralization of energy resources allows new energy sources to enter the distribution network, most notably renewable sources like wind and solar, which not only produces a cleaner energy matrix but also increases competition among energy suppliers, pushing down prices as the number of providers increases. In addition, decentralization allows communities and consumers to exert more control over the energy sources they consume. In some cases, decentralization allows consumers to sell power back to the grid. Advanced distributed energy services can use 5G capabilities to capture real-time systemwide arbitrage opportunities.

As with other sectors, countries should evaluate the value added of 5G infrastructure to support energy applications as compared with other extant technologies. Insight from 5G pilots and tests domestically and abroad among early adopters will be useful sources of information in determining 5G's value proposition, including standalone and non-standalone capabilities within the energy sector.

Transport

The transport industry has entered a period of rapid technological advancement, as demonstrated by the proliferation of electric vehicles; advances in autonomous vehicles; the advent of the ride-sharing economy and digital platforms; advances in big data and machine learning; and rapidly evolving business models, such as e-commerce and mobility as a service, which may cause profound changes throughout the sector. The deployment of 5G mobile broadband has the potential to support and accelerate these developments. Specifically, 5G's features can yield advances in the potential connectivity of vehicles, an increase in the number and ubiquity of connected devices, and improved data availability for transport operations and management.

Three key use cases that may benefit from these advances are connected and autonomous vehicles, smart and efficient logistics, and with the implementation of mobility-as-a-service platforms, improved urban mobility and public transportation. However, 5G's viability, feasibility, and value added are still being assessed as of 2023, and more evidence is needed to demonstrate and assess its impact.

Many applications that can leverage 5G for the transport sector require extensive geographic 5G coverage. Thus, only a few low- and middle-income countries will benefit in the short term, such as China and certain countries in Eastern Europe, Latin America, and Southeast Asia. These developing countries, like developed markets, will face challenges and risks in building a 5G-enabled transport sector, including cybersecurity risks, limited urban public space, and shortages of workers with the required skills.

For limited urban public space, while the rise of 5G and increasingly autonomous private cars risks undermining public transport, better management of road space and greater fluidity of traffic can boost the use of shared autonomous automobiles. At the same time, remote work is cutting commutes, depriving transit systems of customers. Although the related gains are neither negligible nor negative, increased reliance on private vehicles could threaten public transport ridership, increase congestion in city centers, and expand the need for space to park or garage private cars, which raises concerns about sustainability, access for the poor population, and safety. However, improved urban transport based on real-time schedule information may improve ridership.

How—and whether—developing countries prepare themselves for 5G may affect how their transport sectors develop. However, the long lifespan of vehicles and the tendency for older vehicles to remain on the roads longer in developing countries suggests that reaching a critical mass of connected vehicles in these countries may be decades away. Consequently, the need for extensive coverage, long time horizons, and the suitability of other network technologies to support many popular smart transportation applications (such as over-the-air updates and real-time navigation) further suggests that transportation may not be the first area in which 5G is applied commercially, particularly in developing countries. This issue is further compounded by the current state of the evidence on 5G's value added, in which use cases and applications are still being tested. However, the intersection of transport and telecommunications will strengthen the links among the sectors and incentivize public-private partnerships for network ownership and operation, thereby heightening the impacts of 5G applications.

Agriculture

5G and associated technologies have the potential to transform the agricultural value chain from planting and processing to distribution and consumption. Rapid developments in the IoT and cloud computing are driving the development of "smart farming." While precision agriculture considers in-field variability, smart farming goes farther, basing management tasks not only on location but also on data enhanced by contextual awareness and triggered by real-time events. The result should be increased production, lower costs, faster innovation, greater processing power, better monitoring of food quality, and reductions in waste through better market information and logistics (World Bank 2019).

Smart farming practices are already developing without 5G in IoT configurations using 3G, 4G, or NB-IoT and Long-Term Evolution networks. 5G networks can provide additional capabilities to this layer of existing network connectivity, enabling faster data rates to support applications where faster speeds are needed, such as remote veterinary services delivered via high-definition video. 5G will continue to be essential to turn data into effective short-term and long-term decision-making and forecasts.

By 2025, a smart farm might generate millions of farming-related data points daily in addition to millions of additional data points for nonfarm activities, logistics, transportation, and consumption. With 5G, real-time, high-volume data exchange, data processing, and analytics become more feasible, creating possibilities for decision-making at the farm level and for policy making at the macro level. The capabilities of 5G networks will help enable or expand the capacity needed for each of these tasks, as shown in table 2.3.

Mobile operators are increasingly using their networks to test innovative solutions for agriculture (Xi 2017). Examples include monitoring banana production in Colombia to help farmers to deal with flooding, soil oxygen exhaustion, humidity, and low temperatures, thereby boosting productivity by 15 percent; automating irrigation in Spain to reduce water and energy usage, increasing profits by 25 percent; monitoring fish in Viet Nam to reduce fish mortality by 40–50 percent; and the potential for IoT-connected cows worldwide to enable farmers to understand cows' natural cycles, forecast when they are most likely to become pregnant, and predict sickness (Xi 2017).

Current methods of agricultural production are unsustainable in many developing countries. Globalization; climate change; a shift from a fuel-based to a bio-based economy; and competing claims on land, fresh water, and labor will complicate efforts to feed the world without further pollution and resource depletion. Food security is a major issue that will become more urgent in the coming decades, owing to the expected increase in the world population and growing purchasing power in emerging economies. These challenges are related to SDG 2 (zero hunger), SDG 13 (climate action), and SDG 15 (life on land).

Digitally enabled agricultural technologies based on existing 3G and 4G networks already are addressing some key challenges to agriculture in the developing world, including low productivity, lack of market linkages between farmers and buyers, farmers' limited access to financial services, and lack of data to support informed and real-time decision-making. 5G's potential for agriculture in developing countries will vary by country, depending on its context and circumstances. Adoption of 5G is likely to be phased, with

TABLE 2.3 **5G and other technologies in the agriculture sector**

	4G (LoRa, NB-IoT)	5G	5G enabled benefits for IoT in agri-food	IoT application in agri-food
Monitoring	Small size data exchange possible (low internet traffic)	Real-time low or higher volume data exchange (video)	Edge computing enables cloud storage and governance closer to the farmer with lower latency, although 4G is sufficient for this management stage	Drones, video surveillance
			Real-time information about product location, origin, and certification could be unlocked to the end of value chains to better inform consumers	
Sensoring and analysis	Fast, large data exchange difficult	Real-time high volume data exchange. Context- and situation awareness, triggered by real-time events.	Crop monitoring systems; user interfaces (mobile apps, virtual, and augmented reality); early warning in case of food incidents; rescheduling in case of unexpected food quality deviations; and simulation of product quality based on ambient conditions	Drones, sensors, data analytics, AI
Control	Control phase not efficient	Real-time high volume data exchange, edge computing. Autonomous context-awareness by all kinds of sensors and built-in intelligence capable of executing autonomous actions.	Remote controls for irrigation systems, fertilizer systems, climate controllers, and harvesting systems (Miranda et al. 2019). Autonomous tractors and harvesters. Final stage of the management cycle and smart control would be more feasible when 5G connectivity is available, such as intelligent greenhouses, remote, and autonomous farming.	Drones, robots, data analytics, AI, driverless trucks and tractors

Source: Original table for this book.
Note: 4G = fourth-generation mobile network technologies; 5G = fifth-generation mobile network technologies; AI = artificial intelligence; IoT = internet of things; LoRa = long range; NB-IoT = NarrowBand–internet of things.

timing based on the maturity of the driving factors and the challenges posed by existing barriers, as noted in table 2.4.

In the near term, the potential of 4G-enabled IoT appears more appropriate for large-scale farming, but moving toward 2030, certain supply-side drivers of technology will support 5G-enabled agriculture. For example, prices for mobile handsets, telecom and mobile data, sensors, solar panels, satellites, radar systems, and drones have all decreased over the past decade, even as their capabilities have grown. As with other sectors described in this chapter, 5G's main role in agriculture will be to provide an additional layer of advanced connectivity to support new applications that can help resolve the agricultural challenges facing developing countries.

However, for now, most agricultural applications use sensors with relatively low data rates, which can be connected using NB-IoT or other technologies.

TABLE 2.4 **Status of 5G-enabled agriculture as of 2020**

	5G implementation available (mostly in developed countries)	IoT implementation available, without 5G (mostly in middle-income countries)	No smart farming or food production applied (mostly in low-income countries)
5G drivers or barriers in agriculture Digital connectivity Availability and affordability of devices (smartphones and IoT) Adoption of advances in data analytics and exchange (big data, artificial intelligence, and advances in analytics)	All three drivers of 5G are working in favor	Some drivers of 5G are working in favor; some are barriers	All drivers are acting as barriers to 5G
Enablers Innovation ecosystem (especially in finance) A resilient agri-food system capable of leveraging both analogue and digital technologies Human capital Infrastructure	All four enablers are supporting 5G	Some enablers are supportive; some are not	No enablers are functioning to support 5G
Approximate time frame	5 years	10 years	15 years

Source: World Bank 2020c.
Note: 5G = fifth generation mobile network technologies; IoT = Internet of Things.

Therefore, 5G adoption and deployment should be considered carefully along with other technologies for delivering the performance requirements. As with the smart cities, energy, and transport sectors, early adopters must track and evaluate further testing and pilots of new applications and use cases in the agriculture sector.

CONCLUSION

This chapter highlighted 5G's potential role in Industry 4.0 and in achieving the SDGs. The chapter also aimed to help policy makers and other key stakeholders in developing countries to better understand how 5G can strengthen broader mobile network connectivity in support of advances in smart cities, energy, transport, agriculture, and other sectors.

Because of its low latency, high speed, and capacity for connecting many IoT devices, 5G has the potential to support a variety of use cases with varying requirements. For smart cities, 5G can accelerate their growth through sophisticated management of integrated urban systems and services. In the energy sector, 5G can enhance the ability to balance supply and demand, accommodate variable renewable resources, and increase demand-side participation in the production and consumption of electricity and other types of energy. In the transport sector, 5G can support fully autonomous vehicles and intelligent transport systems over the long term. In agriculture—an important sector for low- and middle-income economies—5G can support the shift to a "precision" model in which real-time, in-field optimization enhances productivity and reduces losses and waste.

5G adoption in these sectors is expected not only to yield economic benefits but also to help countries make more rapid progress toward the SDGs. That said, much can be delivered with 4G and other existing solutions, such as NB-IoT, Wi-Fi, and wireless smart ubiquitous networks, in these verticals. Therefore, more time is needed to develop an evidence base that emphatically demonstrates the value-adding role of 5G connectivity. In addition, 5G-enabled sectors carry new risks that policy makers should consider, ranging from cybersecurity to job security. Therefore, governments in developing countries should carefully evaluate the implications of sectoral applications of 5G within their own country's circumstances. Chapter 3 reviews some of the frequently cited risks of 5G.

BIBLIOGRAPHY

3GPP. 2019a. "RP-191048: New Study Item Description on NR-Lite for Industrial Sensors and Wearables." May 27. http://www.3gpp.org/ftp/TSG_RAN/TSG_RAN /TSGR_84/Docs/RP-191048.zip.

3GPP. 2019b. "TS 22.289: Mobile Communication System for Railways; Stage 1." V17.0.0 (Release 17). http://www.3gpp.org/ftp//Specs/archive/22_series/22.289/22289 -h00.zip.

3GPP. 2020a. "TS 22.261: Technical Specification Group Services and System Aspects; Service Requirements for the 5G System; Stage 1." (V17.2.0, Release 17). http:// www.3gpp.org/ftp//Specs/archive/22_series/22.261/22261-h20.zip.

3GPP. 2020b. "TS 28.531: Technical Specification Group Services and System Aspects; Management and Orchestration; Provisioning." (V16.5.0, Release 16). http:// www.3gpp.org/ftp//Specs/archive/28_series/28.531/28531-g50.zip.

5G-PPP (5G Infrastructure Public Private Partnership). 2015. *5G and Energy*. https://5g-ppp.eu/wp-content/uploads/2014/02/5G-PPP-White_Paper-on-Energy -Vertical-Sector.pdf.

5G-PPP (5G Infrastructure Public Private Partnership). 2018a. *Remote Monitoring and Maintenance of Facilities, Infrastructure and Personnel.* https://global5g .org/verticals/5g-energy/usecases/remote-monitoring-and-maintenance -facilities-infrastructure-personnel.

5G-PPP (5G Infrastructure Public Private Partnership). 2018b. *Advanced/Smart Consumption Metering.* https://global5g.org/verticals/5g-energy/usecases /advancedsmart-consumption-metering.

5G-PPP (5G Infrastructure Public Private Partnership). 2018c. *Grid Asset Management and Protection.* https://global5g.org/verticals/5g-energy/usecases/grid-asset-manage ment-protection.

5GCity. 2019. "Deliverable 2.1: 5GCity System Requirements and Use Cases." https://zenodo.org/record/2558210/files/D2.1%205GCity%20System%20 Requirements%20and%20Use%20Cases%20v1.1-CLEAN.pdf.

ACEA (European Automobile Manufacturers Association). 2019. "How Can Automated and Connected Vehicles Improve Road Safety?" *Road Safety Facts.* https://roadsafetyfacts.eu/how-can-automated-and-connected-vehicles-improve -road-safety/.

AgFunder. 2019. "Agri-FoodTech Investing Report: 2019 Year in Review." https:// agfunder.com/research/agfunder-agrifood-tech-investing-report-2019/.

Andrews, Dan, Giuseppe Nicoletti, and Christina Timiliotis. 2018. "Digital Technology Diffusion: A Matter of Capabilities, Incentives or Both?" OECD Economics Department, Working Paper No. 1476, November. http://www .oecd.org/officialdocuments/publicdisplaydocumentpdf/?cote=ECO/WKP% 282018%29246docLanguage=En.

ARENA (Australian Renewable Energy Agency). 2018. "What Are Distributed Energy Resources and How Do they Work?" March 15. https://arena.gov.au/blog/what -are-distributed-energy-resources/.

Barton, James. 2019. "Closing the Generation Gap: Will 5G Be Adopted Sooner in Emerging Markets?" *Developing Telecoms.* October 21. https://www.developingtele coms.com/telecom-business/telecom-trends-forecasts/8864-closing-the -generation-gap-will-5g-be-adopted-sooner-in-emerging-markets.html.

Bass, Frank M. 1969. "A New Product Growth Model for Consumer Durables." *Management Science* 15 (5): 215–27. https://doi.org/10.1287/mnsc.15.5.215.

Beheshti, Babak. 2019. "What 5G Means for Smart Cities." *Smart Cities World,* October 23. https://www.smartcitiesworld.net/opinions/opinions/what-5g-means -for-smart-cities.

BloombergNEF. 2018. *AI Beyond the Buzzword: Energy Applications and Challenges.* New York: Bloomberg Finance LP.

Booij, Johan, Mireille van Hilten, Jeehye Kim, Akanksha Luthra, Krijn Poppe, Parmesh Shah, and Sjaak Wolfert. 2021. *5G and Agriculture.* Washington, DC: World Bank.

Bright, Julian. 2019. *Fixed-Wireless Access Drives Broadband Development in Sub-Saharan Africa.* Ovum. https://www-file.huawei.com/-/media/corporate/local-site/za/pdf /fixed-wireless-access-drives-broadband-development-in-sub-saharan-africa.pdf.

Candell, Richard, Mohamed T. Hany, Kang B. Lee, Yongkang Liu, Jeanne T. Quimby, and Catherine A. Remley. 2018. *Guide to Industrial Wireless Systems Deployments.* NIST Advanced Manufacturing Series 300-4. Gaithersburg, MD: U.S. Department of Commerce, National Institute of Standards and Technology. https://nvlpubs.nist .gov/nistpubs/ams/NIST.AMS.300-4.pdf.

Colle, Serge, and Paul Micalief. 2019. "How DSOs Can Keep Pace with the Fast-Moving Energy Transition." Ernst & Young, March 18. https://www.ey.com /en_gl/power-utilities/how-dsos-can-keep-pace-with-the-fast-moving-energy -transition.

EPRI (Electric Power Research Institute). 2011. *Estimating the Costs and Benefits of the Smart Grid.* https://www.smartgrid.gov/files/Estimating_Costs_Benefits_Smart _Grid_Preliminary_Estimate_In_201103.pdf.

European 5G Observatory. 2020. *5G Cities.* https://5gobservatory.eu/info-deployments /5g-cities/.

Fan, Neng, and Young-Jun Son. 2020. *Smart Grid Optimization.* Institute for Energy Solutions, University of Arizona. https://energy.arizona.edu/research /energy-systems/smart-grid-optimization.

FCC (Federal Communications Commission). 2020. "FCC Proposes the 5G Fund for Rural America." Press Release, April 23. https://docs.fcc.gov/public/attachments /DOC-363946A1.pdf.

Forge, Simon, Robert Horvitz, and Colin Blackman. 2014. *Is Commercial Cellular Suitable for Mission Critical Broadband?* https://op.europa.eu/en/publication -detail/-/publication/246bc6ec-6251-40cb-aab6-748ae316e56d.

Gartner. 2019. "Gartner Says Worldwide IaaS Public Cloud Services Market Grew 31.3% in 2018." Press Release, July 29. https://www.gartner.com /en/newsroom/press-releases/2019-07-29-gartner-says-worldwide-iaas -public-cloud-services-market-grew-31point3-percent-in-2018.

Ghadialy, Zahid. 2020. "Private Networks & 5G Non-Public Networks (NPNs)." Presented at a Workshop on Introduction to Private 4G and 5G Networks (via 3G-4G Blog), February. https://blog.3g4g.co.uk/2020/03/5g-private-and-non-public -network-npn.html.

Grijpink, Ferry, Eric Kutcher, Alexandre Ménard, Sree Ramaswamy, Davide Schiavotto, James Manyika, Michael Chui, and Rob Hamill. 2020. "Connected World: An Evolution in Connectivity Beyond the 5G Revolution." Discussion Paper, February, McKinsey & Co. https://www.mckinsey.com/~/media/McKinsey/Industries /Technology%20Media%20and%20Telecommunications/Telecommunications /Our%20Insights/Connected%20world%20An%20evolution%20in%20 connectivity%20beyond%20the%205G%20revolution/MGI_Connected-World _Discussion-paper_February-2020.ashx.

GSA (Global Mobile Suppliers Association). 2020. "Fixed Wireless Access Global Status Update." May 19. https://gsacom.com/paper/fixed-wireless-access -global-status-update-april-2020/.

GSM Association. 2018. *Network Slicing Use Case Requirements*. April. https://www.gsma .com/futurenetworks/wp-content/uploads/2020/01/2.0_Network-Slicing-Use-Case -Requirements-1.pdf.

GSM Association. 2019. *The 5G Guide*. April. https://www.gsma.com/wp-content /uploads/2019/04/The-5G-Guide_GSMA_2019_04_29_compressed.pdf.

GSM Association. 2020. *Mobile Networks for Industry Verticals: Spectrum Best Practice*. Public Policy Position. May. https://www.gsma.com/spectrum/wp-content /uploads/2020/05/Mobile-Networks-for-Industry-Verticals.pdf.

IHS Markit. 2019. *The 5G Economy*. November. https://www.qualcomm.com/media /documents/files/ihs-5g-economic-impact-study-2019.pdf.

Iji, Matthew, Alla Shabelnikova, Peter Boyland, Dennisa Nichiforov-Chuang, and Lewis Michaelwaite. 2020. "Global 5G Landscape, Q1 2020." *GSMA Intelligence*, April. https://data.gsmaintelligence.com/research/research/research-2020/global -5g-landscape-q1-2020.

Kim, Jae-Hyun, Junsu Kim, and Howon Lee. 2020a. "Toward Future Smart City with 5G." Unpublished manuscript.

Kim, Jeehye, Parmesh Shah, Joanne Catherine Gaskell, Ashesh Prasann, and Akanksha Luthra. 2020b. *Scaling Up Disruptive Agricultural Technologies in Africa*. International Development in Focus. Washington, DC: World Bank. https://doi .org/10.1596/978-1-4648-1522-5.

KPMG International. 2019. "Unlocking the Benefits of 5G for Enterprise Customers: What Telecom Executives Should Know to Take Advantage of the US$4.3 Trillion in Unrealized Value." https://assets.kpmg.com/content/dam/kpmg/cn/pdf/en /2019/04/making-5g-a-reality.pdf.

Kurschner, Dan, and Susan Daffron. 2019. "Farms, Food and 5G." In *5G in Emerging Markets*. Developing Telecoms, October. https://www.developingtelecoms.com /images/reports/5g-em-report-final-1023-4.pdf.

Lanctot, Roger. 2017. *Accelerating the Future: The Economic Impact of the Emerging Passenger Economy*. Strategy Analytics (for Intel), June. https://newsroom.intel .com/newsroom/wp-content/uploads/sites/11/2017/05/passenger-economy.pdf.

Lee, Paul, Mark Casey, and Craig Wigginton. 2019. "Private 5G Networks: Enterprise Untethered." *Deloitte Insights*. https://www2.deloitte.com/us/en/insights/industry /technology/technology-media-and-telecom-predictions/2020/private-5g -networks.html.

Li, Xiaomin, Di Li, Jiafu Wan, Athanasios V. Vasilakos, Chin-Feng Lai, and Shi-Yong Wang. 2017. "A Review of Industrial Wireless Networks in the Context of Industry 4.0." *Wireless Networks* 23 (1): 23–41. https://doi.org/10.1007/s11276-015-1133-7.

Lohmar, Thorsten, Ali Zaidi, Håkan Olofsson, and Christer Boberg. 2019. "Driving Transformation in the Automotive and Road Transport Ecosystem with 5G." *Ericsson Technology Review*, September 13. https://www.ericsson.com/en/reports-and -papers/ericsson-technology-review/articles/transforming-transportation-with-5g.

Maddox, Teena. 2019. "How 5G Will Make Smart Cities a Reality." *ZDNet*, February 1. https://www.zdnet.com/article/how-5g-will-make-smart-cities-a-reality/.

Marek, Sue. 2020. "The Security Conundrum of Network Slicing." *Light Reading*, April 13. https://www.lightreading.com/security/the-security-conundrum-of -network-slicing/d/d-id/758814.

McKetta, Isla. 2020. "Massive Expansions and Huge Improvements in Speed: The Worldwide Growth of 5G in 2020." *Ookla Speedtest Intelligence Blog*, December 11. https://www.speedtest.net/insights/blog/world-5g-report-2020/.

Mendis, Kalpanie, Poul E. Heegaard, and Katina Kralevska. 2019. "5G Network Slicing for Smart Distribution Grid Operations." In 25th International Conference and Exhibition on Electricity Distribution (CIRED), Madrid, June. https://www .cired-repository.org/bitstream/handle/20.500.12455/658/CIRED%202019%20 -%201952.pdf.

Miranda, Jhonattan, Pedro Ponce, Arturo Molina, and Paul Wright. 2019. "Sensing, Smart and Sustainable Technologies for Agri-Food 4.0." *Computers in Industry*, 108: 21–36. https://doi.org/10.1016/j.compind.2019.02.002.

Monserrat, Jose F., Adam Diehl, Carlos Bellas Lamas, and Sara Sultan. 2021. "Envisioning 5G Enabled Transport." Unpublished manuscript.

Moorhead, Patrick. 2015. "Can Jabil Revolutionize the Supply Chain?" *Moor Insights and Strategy*, July 20. https://www.jabil.com/dam/jcr:66273b14-209e-4ed5-8938 -6fcf71049f82/can-jabil-revolutionize-the-supply-chain.pdf.

NEC (Nippon Electric Company). 2020. *What Will 5G Bring to the Smart City?* https:// www.nec.com/en/global/insights/article/2020022501/index.html.

NHTSA (National Highway Traffic Safety Administration). 2015. *Crash Stats*. February. https://crashstats.nhtsa.dot.gov/Api/Public/ViewPublication/812115.

Oliver, Dan. 2020. "Private 5G Network Market Worth $920m in 2020." *5G Radar*, June 2. https://www.5gradar.com/news/private-5g-network-market-worth-dollar 920m-in-2020.

Ranger, Kulveer, ed. 2018. *Digital Vision for Farming*. Atos. https://atos.net/wp-content /uploads/2018/09/DVfFarming-opinion-paper-new.pdf.

Senkhane, Mpinane. 2016. "African Cities Vie for 'Tech Hub of Africa' Throne." *African Cities*, UCLG-A, 5/2016. 56. https://knowledge.uclga.org/IMG/pdf/african _cities_05_en.pdf.

SmartAgriHubs. 2020. "SmartAgriHubs for Africa: Call for Support to DIH [Digital Innovation Hubs]." https://www.smartagrihubs.eu/latest-news/sah-4-africa-call -for-support-dih.

SNS Telecom and IT. 2019. *The Private LTE and 5G Network Ecosystem 2020–2030: Opportunities, Challenges, Strategies, Industry Verticals and Forecasts*. October. https:// www.snstelecom.com/private-lte.

SoftBank News. 2020. "The Future of Logistics: SoftBank Working to Evolve Truck Platooning with 5G." July 1. https://www.softbank.jp/en/sbnews/entry/20200 306_01.

Son, Harrison. 2019. "7 Deployment Scenarios of Private 5G Networks." *NetManias*, October 21. https://www.netmanias.com/en/post/blog/14500/5g-edge-kt-sk -telecom/7-deployment-scenarios-of-private-5g-networks.

Task Force on Transport and Connectivity. 2020. *Towards an Enhanced Africa-EU Cooperation on Transport and Connectivity.* https://citainsp.org/wp-content /uploads/2020/05/20200219_africa-europe_alliance_transport_and_connectivity _taskforce_final.pdf.

Trego, Linda. 2019. "More than 11 Million Vehicles Will Be Equipped with V2X Communications in 2024." *Autonomous Vehicle Technology,* May 17. https://www .autonomousvehicletech.com/articles/1762-more-than-11-million-vehicles-will -be-equipped-with-v2x-communications-in-2024.

UK Government. 2020. "New £65 Million Package for 5G Trials." Press Release, February 20. https://www.gov.uk/government/news/new-65-million-package-for-5g-trials.

UN (United Nations). 2018. "68% of World Population Projected to Live in Urban Areas by 2050, Says UN." News, Department of Economic and Social Affairs. https://www.un.org/development/desa/en/news/population/2018-revision-of -world-urbanization-prospects.html.

UNCTAD (United Nations Conference on Trade and Development). 2019. *Structural Transformation, Industry 4.0 and Inequality: Science, Technology and Innovation Policy Challenges.* Trade and Development Board, Eleventh session, Geneva, November 11–15. https://unctad.org/system/files/official-document/ciid43_en.pdf.

UNDP (United Nations Development Programme). 2020. *Goal 11: Sustainable Cities and Communities.* https://www.undp.org/content/undp/en/home/sustainable -development-goals/goal-11-sustainable-cities-and-communities.html.

Usman, Muhammad, Muhammad Rizwan Asghar, Fabrizio Granelli, and Khalid Qaraqe. 2018. "Integrating Smart City Applications in 5G Networks." In Proceedings of the 2nd International Conference on Future Networks and Distributed Systems, June 26, 2018, 1–5. Amman: Jordan.

UTC (Utilities Technology Council). 2017. *Why Do Utilities Need Access to Spectrum?* https://utc.org/wp-content/uploads/2018/02/Why_Do_Utilities_Need_Access_To _SpectrumOCT-2017FINAL.pdf.

UTC (Utilities Technology Council). 2019. *Cutting through the Hype: 5G and Its Potential Impacts on Electric Utilities.* March. https://utc.org/wp-content/uploads/2019/03 /Cutting_through_the_Hype_Utilities_5G-2.pdf.

Verdouw, Cor, Jacques Wolfert, and Bedir Tekinerdogan. 2016. "Internet of Things in Agriculture." In *CAB Reviews: Perspectives in Agriculture, Veterinary Science, Nutrition and Natural Resources,* Vol. 11. Article 35. https://doi.org/10.1079/PAVSNNR201611035.

WHO (World Health Organization). 2020. *Road Traffic Injuries.* February 7. https:// www.who.int/news-room/fact-sheets/detail/road-traffic-injuries.

Wilson, Stephen. 2020. "Early 5G Fixed-Wireless Access Retail Offers Have Yet to Truly Disrupt the Fixed Broadband Market." Analysys Mason, January 21. https:// www.analysysmason.com/research/content/comments/5g-fwa-retail-rdmb0/.

Wolfert, Sjaak, Daan Goense, and Claus Aage Grøn Sørensen. 2014. "A Future Internet Collaboration Platform for Safe and Healthy Food from Farm to Fork." In 2014 Annual SRII Global Conference (IEEE Computer Society), 266–73. https:// ieeexplore.ieee.org/document/6879694/.

World Bank. 2015. *Smart Cities.* https://www.worldbank.org/en/topic/digital development/brief/smart-cities.

World Bank. 2018. *Innovative Business Models for Expanding Fiber-Optic Networks and Closing the Access Gaps.* Digital Development Partnership, December. http:// documents.worldbank.org/curated/en/674601544534500678/pdf/Main-Report .pdf.

World Bank. 2019. *Future of Food: Harnessing Digital Technologies to Improve Food System Outcomes.* Washington, DC: World Bank. https://openknowledge.worldbank .org/handle/10986/31565 License: CC BY 3.0 IGO.

World Bank. 2020a. *Energy Overview.* https://www.worldbank.org/en/topic/energy /overview.

World Bank. 2020b. *Agriculture and Food.* https://www.worldbank.org/en/topic /agriculture/overview.

World Bank. 2020c. *Agriculture and 5G.* Agriculture Global Practice. Washington, DC: World Bank.

World Economic Forum. 2017. *The Future of Electricity: New Technologies Transforming the Grid Edge.* http://www3.weforum.org/docs/WEF_Future_of_Electricity_2017.pdf.

World Economic Forum. 2020. "The Impact of 5G: Creating New Value across Industries and Society." White Paper, in collaboration with PwC. https://www.pwc.com/gx /en/about-pwc/contribution-to-debate/wef-the-impact-of-fiveg-report.pdf.

Xi, Cipher. 2017. "Feeding the World with Connected Farming." *WinWin Magazine.* https://www.huawei.com/en/publications/winwin-magazine/plus-intelligence /feeding-the-world-connected-farming.

Xinhuanet. 2019. 曾剑秋：5G是国家信息化战略的延伸 会创造出更多需求 *[Zeng Jianqiu: 5G Is an Extension of the National Informatization Strategy That Will Create More Demand],* edited by Deng Cong. http://www.xinhuanet.com/info/2019-04/06/c_137954518.htm.

Ye, Wendy. 2020. "Chinese Agriculture Drone Makers See Demand Rise Amid Coronavirus Outbreak." *CNBC,* March 9. https://www.cnbc.com/2020/03/10 /chinese-agriculture-drone-makers-see-demand-rise-amid-coronavirus-outbreak .html.

Yousaf, Faqir Zarrar, Michael Bredel, Sibylle Schaller, and Fabian Schneider. 2017. "NFV and SDN: Key Technology Enablers for 5G Networks." *IEEE Journal on Selected Areas in Communications* 35 (11): 2468–78. https://doi.org/10.1109/JSAC.2017.2760418.

Zhang, Zoey. 2020. "COVID-19 Catalyzes Commercial Use of 5G in China." *China Briefing,* May 26. https://www.china-briefing.com/news/covid-19-china-5g -commercial-use-which-industries-benefit/.

3

Managing the Risks of 5G Networks

KEY MESSAGES

- The most significant nonfinancial risks of fifth-generation (5G) mobile network technologies involve cybersecurity, sustainability, and health.

- 5G is the first cellular network system planned according to the principles of "security by design." 5G standards resolve many of the security weaknesses of previous mobile generations.

- Although 5G makes much greater use of encryption, authentication, and protection of subscriber identities than previous generations, attacks on the network infrastructure and its subscribers can still be launched from older networks.

- More data channels, greater reliance on configuration by software, and dependence on cloud service providers also present new vulnerabilities in the 5G era.

- The migration to 5G and subsequent winding down of second-generation (2G) mobile network technologies and third-generation (3G) mobile network technologies services have the potential to reduce the overall power consumption and carbon footprint of the information and communication technology (ICT) sector. The network energy efficiency over fourth-generation (4G) mobile network technologies promised by 5G's target capabilities will lead to a further reduction in energy usage. However, the increased data consumption anticipated with 5G adoption and lower data costs over time—along with an increase in the number of 5G-enabled handsets and internet of things (IoT) devices per user—will increase greenhouse gas emissions. The net impact of 5G adoption from these various forces remains difficult to predict—particularly as innovations in technology and applications are advancing rapidly.

- Increased e-waste from pre-5G handsets, related peripherals, and discarded IoT devices remains a concern, particularly in the African and Asian developing countries that receive the bulk of global e-waste.

- Radiofrequency exposure from 5G networks is a growing concern among citizens in many countries, but as of 2023, no conclusive evidence of risks to people from such exposure has been found. Governments must manage communications around this topic and refer citizens to trusted institutions producing the latest research and guidance on health concerns.

INTRODUCTION

So far, this book has presented the design principles of 5G technology that distinguish it from previous generations of mobile wireless technologies, has discussed the implications for digital divides, and has described the array of potential benefits that 5G networks can bring to a few notable sectors related to the Sustainable Development Goals. This chapter discusses some major risks that governments must consider, namely those for cybersecurity, including risks to various user groups; sustainability, including climate change; and health, including risks from electromagnetic fields, and concludes with a discussion of risk management.

CYBERSECURITY RISKS

Chapter 2 of this book reviewed the anticipated benefits of using 5G networks to develop critical sectors of the economy. There is a growing understanding that 5G networks may evolve into essential parts of the supply chain for many critical services and applications. In such cases, confidentiality and privacy will not only be necessary, but the integrity and availability of networks also will become major national security concerns and challenges (NIS Cooperation Group 2019). Developing countries seeking a well-connected future must prioritize securing 5G networks in ways that will serve a range of sectors, applications, and use cases.

All generations of mobile communications technology are prone to attack. In 2019, a mobile network could expect over 3,000 attack attempts per day on average (Positive Technologies 2019). Common forms of attacks include location tracking, eavesdropping (for corporate or state-sponsored espionage), Short Message Service theft (for example, to breach the two-factor authentication widely used in electronic payment systems), and data theft (for any purpose). Although such attacks are harmful in any country, they pose a bigger risk to developing countries, where cybersecurity capacity and awareness of the dangers are limited (Amin et al. 2020).

New 5G technology promises better protection from its incorporation of "security by design." During the 5G standardization process, the security and privacy of future users were carefully considered to solve problems identified in previous mobile network generations. Consequently, 5G is more secure than its predecessors. The significant improvements incorporated into the 5G standards include stronger encryption of data and better verification of network users' identities.

5G also integrates new technologies, such as the software-defined network, network virtualization, and multi-access edge computing. However, its complexity in design and operation contributes to new vulnerabilities that are not yet fully understood. Some threats are traceable not to 5G technology per se but to new network attachments in the form of IoT devices (Kuchler 2017). Likewise, the evolving nature of cybercrime poses new risks. In addition, because the less secure networks of previous generations are still in use, attacks utilizing those networks can still reach 5G through the interconnections required for roaming. The reuse of core elements of 4G technology in 5G non-standalone deployments also prevents the full implementation of the unique security advances available in 5G standalone networks. Finally, threats may rise with the use of advanced technologies, such as machine learning, artificial intelligence (AI), and quantum computing, all of which will evolve in parallel with 5G networks, using them, and being used by them, for operation and management.[1]

Enhanced Security Architecture through Security by Design

The 5G security approach—"security by design"—improves on previous cellular generations, which relied on insecure protocols such as Signaling System 7 and Diameter.[2] Security by design is based on three principles. First, like 4G, 5G applies mutual authentication, confirming that the sender and receiver have established trust and their relationship is secured end to end. Second, 5G presumes an open network, which rules out any assumption of safety from overlaid products or processes and consequently necessitates a security-by-design approach from the start. Third, as for previous generations of cellular, 5G mandates the encryption of inter- and intra-network traffic, rendering intercepted information worthless.

The 5G security architecture described in the 3rd Generation Partnership Project's technical specifications 33.401 and 33.501 includes six key areas of improvement over previous network architectures (3GPP 2020a, 2020b):

- *Network access security* protects against attacks on access links. A major enhancement is the introduction of a "subscription concealed identifier" to protect user privacy, a one-time identity that is never repeated and is exchanged between the user and the network only in an encrypted format.

- *Network domain security* pertains to features that enable nodes to protect against attacks and securely exchange signaling data and user data. This concept includes trust models for the authentication of a requesting network and enhanced home network control to validate the legitimacy of a subscriber's service request.

- *User domain security* focuses on the security of user access to mobile equipment. This concept includes internal security mechanisms, such as personal identification number codes, to ensure security between the mobile equipment and the universal subscriber identity module.

- *Applications* and the features that enable applications in the user and provider domains are used for securely exchanging messages.

- *Service-based architecture* and its security and features enable it to communicate securely within the serving network domain and with other network domains. Such features include the registration, discovery, and authorization of network functions, as well as protection for service-based interfaces.

- *Visibility and configurability of security* enable users to be informed whether a security feature is in operation and whether services should depend on the security feature. This design feature is critical, as many of the security features defined in the 5G specifications are not mandatory for the mobile operator to implement.

Open radio access network (RAN) technology is a network architecture that allows for interoperability between RAN components from different vendors to promote flexibility, innovation, and competition while reducing reliance on a single vendor (vendor lock-in is common in traditional network architecture). Open RAN also has important cybersecurity implications for 5G, given that it is increasingly being used to accelerate 5G adoption through shared network architecture.

Although the "open" aspect may sound alarming from the perspective of cybersecurity, it can support the cybersecurity of 5G in a few important ways.[3] With open RAN, network operators can choose among vendors for different components of the network to ensure the most optimal vendor and solution arrangement. In addition, open RAN can leverage AI and machine learning technology focusing on specific components of the network architecture through interconnected solutions to improve real-time response to threats. Finally, open RAN can be integrated with other security solutions, such as firewalls and intrusion detection systems, to provide more comprehensive security.

5G's Vulnerabilities and Threat Surface

Although the standardized network architecture for 5G offers significant advantages, it also presents new intrinsic vulnerabilities related to software flaws, cloud hosting, and supply chain risks, all of which network operators and governments must consider. Because 5G is a software-based network, the network operation is fully decoupled from the underlying hardware components, instead relying on software for all its functions. This process provides flexibility and improves network economics. However, software flaws can have devastating effects on network operation and availability, especially where core network functions are concerned. Therefore, the prior verification of software and its updates is important.

In addition, 5G networks' readiness for cloud-based hosting creates vulnerabilities that arise from the need to rely on a third party, the cloud service provider. Threats to cloud services could reveal confidential information and affect the availability of the entire 5G infrastructure (ENISA 2019).

Supply chain risks are also widely viewed as national security challenges to the extent that 5G infrastructure permeates the economy. Consequently, increased attention has been placed to securing the 5G supply chain and using ex ante measures—in addition to ex post mitigation measures—to restrict vendors and suppliers profiled as high risk.

Looking ahead, advances in technologies like machine learning, AI, and quantum computing will also affect the 5G security framework. The first two will strengthen and improve network monitoring capabilities, cyberthreat intelligence analytics, malware analysis, and detection of anomalies. AI-based solutions may prove efficient for confronting certain types of attacks, and automation can help improve threat identification and mitigation. However, increased automation of network management will make it more susceptible to "adversarial AI" attacks that seek to interfere with automated decision-making.

Quantum communications, information processing, and computing are quickly evolving scientific fields with implications for 5G network security. As technologies based on these disciplines enter the commercialization phase, they may have a dramatic effect on the entire ICT ecosystem, including 5G.

As with machine learning and AI, the anticipated impact is expected to be twofold. On the one hand, quantum-based technologies will substantially contribute to overall cybersecurity by providing solutions to enduring security problems. For instance, quantum key distribution will provide a secure way to share encryption keys over an insecure channel, while quantum digital signatures will provide absolute assurance of authenticity, integrity, and nonrepudiation of electronic messages. On the other hand, the extraordinary processing power provided by quantum computing puts all widely adopted cryptography and security algorithms at risk, including those used to protect 5G networks. The 3rd Generation Partnership Project and other standardization bodies are analyzing the impacts of quantum-based capabilities on 5G and considering ways to address the risks (3GPP 2019; ETSI 2015). Future releases of 5G specifications will include an updated definition of quantum-safe cryptography to ensure that the encryption algorithms used on 5G networks are sufficiently resistant to attacks by quantum computers.

Beyond new intrinsic risks and those from advanced technologies, exposure to both accidental and malicious attacks—also known as the "threat surface"[4]—will also increase with 5G, mainly due to the major increase in data traffic and novel types of IoT users.[5] It is estimated that the number of these operatorless terminals (the "things" in IoT) connected to the internet will soar in the coming years (Gartner 2019):

- By 2023, the installed base of 5G endpoints could reach 49 million units.

- By 2024, 5G networks are expected to carry 35 percent of global internet traffic, nearly tripling from 2020 (Peisa et al. 2020), further enlarging the threat surface.

At the same time, attack types common to networks of previous generations will remain relevant to 5G. Such attacks include *botnets* that take control of connected devices and use them as weapons and denial-of-service attacks that flood a network or website with so much traffic that it cannot respond.[6] These and other types of attacks relevant to 5G are summarized in figure 3.1. The security threats stemming from the IoT are not caused by 5G's security technology but instead by the inherent vulnerability of the IoT devices; these threats must be assessed and addressed separately.

Interconnections with pre-5G networks create a separate category of threats. Interconnection implies trust, which is an intrinsic security problem of mobile networks that can be solved only by universal migration to 5G. In 2G technology (Global System for Mobile Communication), 3G (Universal Mobile Telecommunications System), and 4G (Long-Term Evolution), the interconnections between networks are based on insecure signaling protocols. The vulnerabilities of these protocols can be exploited to launch attacks that can compromise users' privacy, with location tracking and eavesdropping among the possibilities (Engel 2014). Unfortunately, 5G networks cannot avoid such interconnections, as they are required for the roaming arrangements that give users universal connectivity for their calls and data.[7] However, the 5G standardization process strengthened the security

FIGURE 3.1 Summary of the 5G threat surface

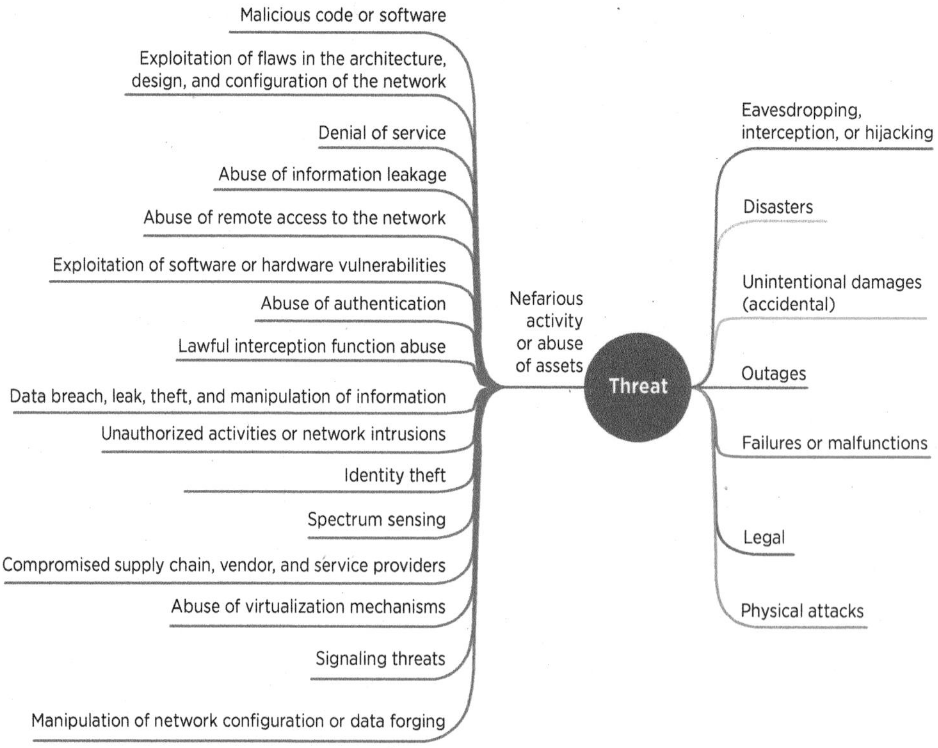

Malicious code or software

Exploitation of flaws in the architecture, design, and configuration of the network

Denial of service

Abuse of information leakage

Abuse of remote access to the network

Exploitation of software or hardware vulnerabilities

Abuse of authentication

Lawful interception function abuse

Data breach, leak, theft, and manipulation of information

Unauthorized activities or network intrusions

Identity theft

Spectrum sensing

Compromised supply chain, vendor, and service providers

Abuse of virtualization mechanisms

Signaling threats

Manipulation of network configuration or data forging

Nefarious activity or abuse of assets

Threat

Eavesdropping, interception, or hijacking

Disasters

Unintentional damages (accidental)

Outages

Failures or malfunctions

Legal

Physical attacks

Source: ENISA 2019.

of internetworks (where two or more computers can interact in a shared system) through a function known as the "security edge protection proxy." While these security measures verify the identity of the external network element, they still depend on trusting that the external network operator will not exploit the connectivity maliciously.

Not all threats have the same impact. The severity of threats varies by network segment, creating a varied risk landscape, which is mapped in table 3.1. Although all 5G network segments are susceptible to threats, threats to the core network, if they materialize, generally have a greater impact.

TABLE 3.1 **5G threat surface, by network segment**

Network segment	Threat
Mobile device and user equipment	• *Mobile to network:* A large number of infected devices controlled by an attacker launch a distributed denial-of-service attack against the network. • *Mobile to the internet:* A large number of infected devices controlled by an attacker launch a distributed denial-of-service attack against an internet site. • *Mobile to mobile:* Infected devices launch an attack against other mobile devices to spread malware. • *Internet to mobile:* Malware is delivered to the mobile network through an infected internet site.
RAN	Most published attacks against RANs involve rogue base stations, also commonly referred to as "cell-site simulators" or "IMSI catchers".[a] The equipment of the targeted user is attacked during its initial attachment to the network procedure or by using the IMSI paging feature (paging attacks). Information about a specific IMSI can be used for other attacks. Although it was believed that improvements in the 5G standards mitigated such attacks, more recent publications (Borgaonkar et al. 2019; Hussain et al. 2019) suggest that they are still possible.
Core network	Because they are based on internet protocols, 5G networks are potentially vulnerable to internet protocol–based attacks. Attacks on the core network (for example, via a distributed denial-of-service attack on 5G core network elements such as the access and mobility management function, the authentication server function, or the unified data function) may provoke loss of coverage over wide areas. Attacks also could provoke significant degradation of network performance.
Network connectivity and signaling	With no authentication built in by design, the signaling protocols used in 2G, 3G, and 4G technologies are insecure and vulnerable to attacks that can compromise user privacy. 5G's security architecture introduces the security edge protection proxy to address that vulnerability. However, this solution is not yet fully standardized.
Roaming network	Given that networks based on previous generations are much less secure than 5G, attacks generated in those networks can reach 5G through the trusted interconnection channels that are required to serve roaming subscribers.
Applications and services	Over-the-top applications and services provided over 5G networks may be subject to attack. With growth in traffic volume and in the number of connected devices on which applications and services rely, the associated risks of attacks also increase.

Source: Fixler and Weinberg 2020.
Note: 2G = second-generation mobile network technologies; 3G = third-generation mobile network technologies; 4G = fourth-generation mobile network technologies; 5G = fifth-generation mobile network technologies; IMSI = international mobile subscriber identity; RAN = radio access network; RBS = radio base station.
a. An RBS, a device operating on spectrum licensed to network operators but not owned or operated by the network operator or the subscribers, acts like a cell tower and broadcasts a signal pretending to be part of a licensed mobile network, seeking to trick an individual user's equipment into connecting to it. The hardware necessary to build an RBS can be obtained using inexpensive off-the-shelf parts. The software required to operate an RBS is open source and freely available (Nakarmi, Kuman, and Norrman 2018; NIST 2017).

5G's Risk Landscape for Different User Groups

The dangers posed by 5G cyberattacks vary by type of user. The five major user groups are individual users (end users), operators of connected IoT devices, mobile network operators or IoT service providers, businesses, and national governments.

- *Individual users* rely on their mobile devices for ready access to services related to their personal and professional lives. The inability to access mobile services or information stored on their devices, the leakage of private information, and the compromises of mobile-based two-factor authentication represent serious risks to individual users, as does the risk of invasion of privacy through location tracking, remote activation of the device's camera or microphone, or impersonation[8] (Rupprecht et al. 2020). The most severe risks to individual users are the potential loss of information due to ransomware and financial harm due to the hijacking of two-factor authentication.

- *Operators of connected IoT devices* face two major risks: the leakage of information collected by the device and the interruption of services. With billions of connected IoT devices, the amount of private or otherwise sensitive information will drastically increase and become at risk of cyberattack. Targeted information may include aspects of a person's life, such as their movements, activities, or health. Information related to critical machine operations may also be targeted. The risk is especially severe in the case of IoT devices providing mission-critical services, such as medical devices and connected cars. The most severe risk is the interruption of services due to denial-of-service attacks.

- *Mobile network operators or IoT service providers* confront three main risks: the nonavailability or diminishment of network services, the corruption of billing functions, and the leakage of consumer information. Curtailment of services may be local, with affected areas limited to the coverage of a specific tower or location. A much more severe risk is the loss of coverage over a wide area or even a national territory, which could result from an attack on core network functions. Interference with the ability to conduct precise billing is a substantial risk to network operators, as is the leakage of consumer personal data, which may result in violation of privacy regulations.

- *Businesses,* with their increased reliance on digitalization, must understand that the cyber risks from 5G will increase for enterprises of all sizes. The majority of cyberattacks are likely to be disruptions of services delivered over mobile channels to consumers, attacks on internal information technology systems through cellular channels, disruptions of web-based internet services, and exposure of the locations and movements of company employees.

- *National governments* face risks that have implications for public safety, including disruption of critical infrastructure services in telecommunications, the water supply, energy, transportation, policing, and health care, making smart cities particularly vulnerable. All risks in this category are assessed as severe, as indicated in table 3.2.

This chapter discusses only the most significant cybersecurity risks of 5G networks. Many threats and vulnerabilities are contingent on national circumstances, and policy makers in developing countries should compile an inventory of all possible 5G security risks, considering the detailed analyses that have been produced elsewhere (for example, refer to CISA 2019a, 2019b; ENISA 2019).

TABLE 3.2 **Summary of potential risks for various 5G user groups**

User group	Potential damage	Parameter			Total risk
		Severity	Effort	Feasibility	
Individual users	Availability of service (denial-of-service attack)	1	1	5	3.0
	Privacy invasion	2	3	4	3.0
	Leakage of private information	2	3	5	3.5
	Location tracking	2	2	5	3.5
	Impersonation	3	4	4	3.5
	Loss of information (ransomware)	3	3	5	4.0
	Financial losses (attack on two-factor authentication)	4	3	5	4.5
Operators of connected IoT devices	Leakage of information	3	4	4	3.5
	Interruption of service	5	3	5	5.0
Mobile network operators or IoT service providers	Leakage of consumer information	4	5	3	3.5
	Loss of local network availability	4	3	4	4.0
	Loss of countrywide availability	5	5	3	4.0
	Disruption of billing	4	3	4	4.0
Businesses	Tracking of corporate employees	2	2	3	2.5
	Disruption of web-based services	2	2	5	3.5
	Loss of access to company resources	3	3	4	3.5
	Interruption of services to consumers	4	3	4	4.0
National governments	Disruption of critical infrastructure	5	4	4	4.5
	Disruption of communications infrastructure	5	4	4	4.5

Source: Fixler and Weinberg 2020.
Note: Risks are scored subjectively from lowest (1) to highest (5). Assessments are based on three parameters. *Severity* pertains to the level of potential damage or disruption, and the score is based on an assessment of the number of people affected and the ensuing damage. *Effort* indicates the level of effort or skill required to implement the attack, and the score is based on the complexity of the implementation of the attack. *Feasibility* is an estimate of the likelihood of a given attack, and the score incorporates the effort score. The *total risk score* is based on the average of the scores for severity and feasibility. IoT = internet of things.

Managing Cybersecurity Risks

5G's softwarization and its interconnections with older mobile networks create vulnerabilities, and its threat surface is further enlarged by the amount of critical business data the networks will carry. Because no one is better placed to maintain security, operators of 5G networks are urged to be proactive in implementing security measures, for example, by establishing internal network security certification; establishing processes to verify supply chains, including vendor diversification and vendor verification for new network elements, software updates, and maintenance processes; and vetting cloud hosting services. The zero-trust principle, which aligns user access and privileges according to preestablished rules, should be adopted when protecting one 5G subsystem from another, with continuous application of security assurance procedures. Mobile operators with IoT offerings will need to execute similar measures, including internal certification processes, supply chain verification, cloud hosting service verification, personnel verification, and increased security services for sensitive users.

The role of governments in defining requirements and supporting 5G security at the national level will be even more critical than under previous generations of mobile communications. Governments should undertake thorough and regular national 5G cybersecurity risk assessments to identify vulnerabilities, threats, and risks and adjust mitigation measures accordingly. Legislatively, governments should define and establish a 5G certification framework that includes minimum security requirements to be implemented in 5G networks, as well as arrangements for implementing certification procedures for various defined network elements.

A national cybersecurity center functioning as the executive agency for cybersecurity policy should be closely involved throughout the process. Certified equipment and procedures should be promoted consistently. To facilitate validation and verification of software updates before implementation, governments may wish to establish a national 5G cybersecurity testbed in collaboration with the private sector. Governments should anticipate the demand for cybersecurity-related skills and ensure an adequately trained labor force.

Because stakeholders' priorities will vary at every stage of the 5G implementation process, some actions must be taken before others. Countries are at varying stages of 5G deployment, with some just starting to debate the costs and benefits of 5G technology, others making concrete plans, and still others scaling up their commercial 5G networks. The following three tables provide recommendations based on 5G status: table 3.3 focuses on countries without a current 5G implementation process, table 3.4 focuses on countries in the planning phase, and table 3.5 focuses on countries already implementing 5G.

TABLE 3.3 Recommendations for countries without a current 5G implementation process

User group	Recommendation
Individual users	• Encourage users to consider mobile needs.
Mobile network operators	• Participate in a national 5G security group (see description under "National governments"). • Define an internal network security certification process. • Define a supply chain verification processes, including vendor diversification, verification of new network elements, software updates, and maintenance processes. • Define the verification processes for cloud hosting services. • Define the personnel verification program that includes periodic random verifications, the distribution of responsibility (avoiding single points of failure in responsibility), and the implementation of security training programs.
Mobile IoT service providers	• Identify the IoT services requiring 5G capabilities for performance and enhanced security.
Businesses	• Build a business case for 5G networks, asking whether to implement enhanced mobile broadband; massive machine-type communications; or ultra-reliable, low-latency communications.

National governments	• Conduct a national 5G cybersecurity risk assessment.
	• Define and adopt a 5G certification framework to include the following:
	  • Scope of application for hardware and software, specifying hardware and software components; software updates, applications, and services; and cloud hosting services
	  • Minimum security requirements for 5G networks, including the type of encryption used and standardized (but not formally mandated by 3GPP, such as the "subscription concealed identifier")
	  • Certification process
	• Create a national 5G cybersecurity group comprised of 5G network operators, operators of critical infrastructures, and the national cybersecurity center.
	• Estimate the future demand for cybersecurity skills, and develop educational tracks to ensure an adequately skilled labor force.

Source: Original table for this book.
Note: 3GPP = 3rd Generation Partnership Project; 5G = fifth-generation mobile network technologies; IoT = internet of things.

TABLE 3.4 Recommendations for countries planning 5G implementations

User group	Recommendation
Individual users	• Identify 5G utilization needs, including security aspects.
Mobile network operators	• Implement an internal network security certification process. • Implement a defined supply chain verification process. • Implement a verification process for cloud hosting services. • Recognize the need to interconnect with legacy networks, verify the implementation of the "security edge protection proxy" per 3GPP requirements, and maintain a list of external connectivity partners. • Because the initial 5G implementation in non-standalone mode will be based on 4G (LTE) cores, include in roaming agreements the definitions of proper interconnectivity security procedures, using diameter end-to-end signaling security solutions as specified in recent updates of GSM Association standards FS.19, FS.21, and IR.88. • Implement a personnel verification program.

(continued)

TABLE 3.4 *(continued)*

User group	Recommendation
Mobile IoT service providers	• Define an internal network security certification process. • Define the supply chain verification process to include vendor diversification and the verification of vendors, new network elements, software updates, and maintenance procedures. • Define the verification process for cloud hosting services. • Define a personnel verification program to include periodic random verifications, the distribution of responsibility (avoid lone responsibility for critical events), and the implementation of security training programs.
Businesses	• Start the commercial phase of security for the network.
National governments	• Set up the implementation of the certification process through public, private, or international certification bodies. Keep the national cybersecurity center closely involved in the definition process. • Encourage the implementation of certified equipment and procedures in 5G. Consider economic incentives such as full or partial compensation for equipment-related customs fees, and review the frequency of licensing fees and loosening of coverage targets for heightened security services. • In collaboration with the private sector, establish a national 5G cybersecurity testbed to validate and verify hardware and software before 5G deployment. • Create a national forum to promote and debate open 5G issues, including cybersecurity.

Source: Original table for this book.
Note: 3GPP = 3rd Generation Partnership Project; 4G = fourth-generation mobile network technologies; 5G = fifth-generation mobile network technologies; IoT = internet of things; LTE = Long-Term Evolution.

TABLE 3.5 Recommendations for countries in the process of implementing 5G

User group	Recommendation
Individual users	• Use only certified equipment. • Put personal precautionary measures into practice. • Consider subscribing to dedicated mobile protection services.
Mobile network operators	• Develop and provide increased protection services for the security of sensitive users (for example, value-added services). Such services could provide improved security from well-known attacks, such as IMSI catchers.
Mobile IoT service providers	• Implement an internal network security certification process. • Implement a supply chain verification process. • Implement a verification process for cloud hosting services. • Implement a personnel verification program. • Develop and provide increased protection services for the security of sensitive users (for example, value-added services). Such services shall provide improved security from well-known attacks, such as IMSI catchers.
Businesses	• Implement a cybersecurity training program. • If relevant, subscribe to increased protection services from 5G network operators and IoT service providers. • Ensure that corporate assets are protected from mobile-sourced cyberattacks.
National governments	• Operate a war room that manages and supports all the above groups against cyberattacks.

Source: Original table for this book.
Note: 5G = fifth-generation mobile network technologies; IMSI = international mobile subscriber identity; IoT = internet of things.

SUSTAINABILITY RISKS: 5G's CARBON FOOTPRINT AND IMPACT ON E-WASTE

The most important sustainability concerns surrounding 5G's deployment center on its impacts on climate change and e-waste, which have important implications for both industry and governments seeking to mitigate negative environmental impacts resulting from a 5G-enabled economy. As of 2022, little research had been done on forecasting an accurate estimation of the net impact of 5G infrastructure and technologies. The research that has been carried out is limited by unknowns for if and when 5G target capabilities will be reached and commercialized, as well as by the difficulties involved with predicting the innovations in technologies and applications. This section explores the drivers of sustainability concerns related to 5G deployment and adoption, focusing on energy consumption and e-waste.

Energy Consumption and E-Waste

Mobile network operators note that the energy consumption of a 5G base station may be two to three times greater than that of a 4G base station (GSM Association 2019). However, if 5G's target capabilities are reached, these technologies would yield significant efficiency improvements over 4G, with up to 100 times greater network energy efficiency as compared with 4G, as noted in chapter 1 of this book. However, the overall balance between efficiency gains and increased energy consumption is difficult to predict, and it will depend on the overall increase in global data consumption, which will continue to rise as the cost of mobile data decreases and the number of subscribers increases.

A similar phenomenon manifests beyond network infrastructure, where end user devices, such as handsets and IoTs, are becoming more energy efficient while also increasing in overall number of devices per user as costs go down and the breadth of innovative applications expands. This reality not only pushes up greenhouse gas emissions but also increases the amount of e-waste, as will be discussed later in this section.

5G adoption may yield additional energy savings due to the reduction in overall electricity consumption derived from winding down earlier generations of network technologies, as indicated in figure 3.2. For an accurate measure of the energy efficiency of 5G networks as it relates to enhancing climate adaptation, more research and better long-term data are needed. Based on actual deployments over time, the findings of this research would enable the improved forecasting of the impact of future energy-saving innovations in manufacturing, infrastructure, and IoT devices and handsets.

The net effects of these various efficiency and consumption trends in 5G technology, deployment, and applications make it difficult to predict accurately the overall impact of a 5G-enabled economy on overall energy consumption and greenhouse gas emissions.

FIGURE 3.2 **Mobile network electricity consumption to 2030**

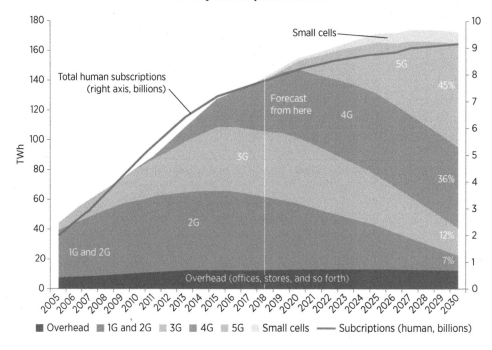

Source: ITU-T 2020.
Note: 1G = first-generation mobile network technologies; 2G = second-generation mobile network technologies;
3G = third-generation mobile network technologies; 4G = fourth-generation mobile network technologies;
5G = fifth-generation mobile network technologies; TWh = terawatt hours.

E-waste is another environmental risk of 5G deployment and adoption. Already a significant problem in the pre-5G world due to technology obsolescence, electronic waste encompasses not only mobile phones but also computers and laptops, televisions and displays, and other electrically powered products that are eventually dumped into landfills (US Environmental Protection Agency n.d.). Each generation of mobile technologies is accompanied by myriad new devices, such as handsets, cables, power blocks, listening devices, and memory storage components. The emergence of 3G networks and 3G-enabled smartphones required consumers to replace devices designed for earlier generations to enjoy the faster speeds and new features of 3G connectivity. Similarly, the emergence of 5G networks will result in the replacement of older-generation phones (particularly 3G and earlier handsets), which cannot communicate with 5G base stations (4G devices can continue to be used with non-standalone 5G networks).

The related e-waste risk is greatest when these older phones are not disposed of correctly, as they contain hazardous materials that can seep from landfills and other waste sites into the soil and groundwater (US Environmental Protection Agency n.d.). The lithium-ion batteries that power many smart phones make up a large share of the hazardous materials found in handsets.

Rapid innovation in connected IoT devices, spurred by 5G technology, may also increase the amount of e-waste from 5G as compared with previous generations of mobile networks. As with handsets, the batteries used to power 5G-enabled IoT devices will amplify the amount of e-waste produced (Knight 2020). Refurbishing and recycling are ways to mitigate this e-waste, along with product design innovations that focus on reducing the amount of hazardous material used in consumer products. Future innovations in waste management may also help resolve this problem, especially if the costs of extracting hazardous materials and recovering rare earth metals can be reduced. Unfortunately, the problem of e-waste from 5G network deployment and adoption is likely to pose an even greater risk to developing countries, particularly those in Africa and Asia, which already receive much of the world's e-waste (Dalul 2020).

Guidelines on Carbon Emissions and E-Waste

In 2019, a group of mobile network operators representing more than two-thirds of all global mobile connections agreed to reveal data on their networks' climate impacts, energy use, and greenhouse gas emissions (GSM Association 2019). The group has expanded to include the Global Enabling Sustainability Initiative, the International Telecommunication Union, and the Science Based Targets Initiative to facilitate a more collaborative effort to understand the carbon footprint of ICT as a whole and to help ICT companies "set targets in line with climate science" (ITU-T 2020, v).[9] This issue has led to the drafting of ITU-T Recommendation L.1470, which outlines how the sector could reduce greenhouse gas emissions by 45 percent by 2030. As of 2022, many operator groups representing almost half of mobile connections and almost two-thirds of industry revenue worldwide, had committed to working toward science-based targets (GSM Association 2022).

For e-waste guidelines, a few multilateral conventions on the transboundary movements of hazardous waste and disposal have been developed, of which the Basel Convention is perhaps the most prominent example. The Basel Convention was designed to address the transfer of hazardous waste from developed to developing countries. Adopted in 1989, the convention focused on the transport of hazardous waste more broadly rather than on the narrower domain of mobile network devices, let alone on 5G handsets, batteries, and IoT devices. In 2019, technical guidelines were adopted under the convention that focused on categorical definitions and transboundary movements of e-waste, as well as on guidance for facilities.[10] Although these guidelines reference the collection, movement, refurbishment, recovery, and recycling of end-of-life mobile phones, they do not address any 5G-inspired updates or revisions, which may be needed in future years to address the inevitable exponential increase in 5G-enabled IoT products and devices.

Managing Sustainability Risks

Environmental risk management requires attention to the energy consumed in the manufacturing and use of mobile equipment and to the growing amount of e-waste, which pose not only environmental dangers but also health hazards. E-waste is a concern for the few developing countries that receive the global e-waste generated in other countries. This problem will undoubtedly grow with new generations of mobile technologies beyond 5G.

Governments that are concerned about the carbon footprint of the 5G era have an opportunity to proactively address the overall carbon footprint of the ICT sector, not merely the latest generation of mobile technologies. Governments can do this by setting and adopting national regulatory standards aligned with international best practices and standards. The Intergovernmental Panel on Climate Change would be a useful starting point for policy makers seeking to develop domestic standards that protect the environment within and beyond their borders (IPCC 2018). Countries may also choose to participate in and adhere to multilateral conventions related to e-waste, such as the Basel Convention.

HEALTH RISKS

Civil society's concerns about the possible health effects of radiofrequency (RF) exposure from 5G's beamforming and network densification intensified after the onset of the COVID-19 crisis in 2020. Unscientific claims have been made that 5G is a vector that spread the novel coronavirus (Grad 2020). Dozens of arson attacks on cell towers and confrontations with maintenance workers and engineers have been reported in Europe, New Zealand, and North America (Waterson and Hern 2020). Developing countries were affected as well, to the point where governments, for example, of Nigeria, had to reassure their citizens that no 5G base stations were deployed ("Nigeria: NCC Clears Doubts Over 5G, COVID-19 and Security" 2020). This issue led the World Health Organization (WHO) to declare that no scientific evidence exists to support the claim that 5G has adverse health effects caused by exposure to electromagnetic fields (EMFs) (WHO n.d.).

The broader discourse linking health risks to 5G networks has three themes, which are discussed in the following sections:

- Radiation health risks of 5G networks
- Radiation health risks from cellular usage, including devices and infrastructure
- Purported link between 5G network deployment and the COVID-19 pandemic.

Research on Radiation Exposure and Cellular Use

Human exposure to EMFs at home and at work has been well documented since the 1960s (Gye and Park 2012). Research has covered mobile networks as well as other communications media, such as Bluetooth and Wi-Fi. Studies have been conducted in academia, industry, and global and national health institutions, and none have demonstrated a direct causal link between cellular phone usage and cancer.

Following a 10-year study from 2008 to 2018, the US Food and Drug Administration concluded that "there is insufficient evidence to support a causal association between radiofrequency radiation exposure and [tumor formation]" (US Food and Drug Administration, 2022). The National Toxicology Program in the United States does not include RF radiation on its list of known or suspected carcinogens. The US Federal Communications Commission (2020) and US Centers for Disease Control and Prevention (2014) have also stated publicly that they have no scientific evidence of a causal link between wireless devices and cancer. Scientific reviews of existing literature have produced no conclusive evidence directly linking exposure to wireless technology to causing adverse health effects (WHO 2020).

The primary risk of RF radiation exposure from current 5G technologies is tissue heating (WHO 2020). However, the high-band allocations for 5G are much less penetrating than the radio waves used by television and FM radio broadcasting, and the energy emitted from 5G radio waves is well within the range of nonionizing radiation, as illustrated in figure 3.3.

The public concern about the health risks of cellular technology may be explained by the absence of full and unanimous consensus among researchers. Although dissent is a common characteristic of research findings

FIGURE 3.3 Radiation across the electromagnetic spectrum

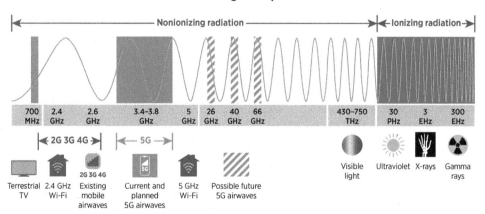

Source: UK Office of Communication n.d.
Note: 2G = second-generation mobile network technologies; 3G = third-generation mobile network technologies; 4G = fourth-generation mobile network technologies; 5G = fifth-generation mobile network technologies; EHz = exahertz; GHz = gigahertz; MHz = megahertz; PHz = petahertz; THz = terahertz.

in most domains, the disparate interpretations of the research findings in this field have aroused public concern in many countries. One example is a statement from WHO's International Agency for Research on Cancer (IARC 2013), which classified RF radiation as a Group 2B agent, a category reserved for items that are "possibly carcinogenic to humans." According to the agency, this label is applied to agents "[for] which there is *limited evidence of carcinogenicity* in humans and less than *sufficient evidence of carcinogenicity* in experimental animals" (IARC 2013, 30). The label may also be used when there is *"inadequate evidence of carcinogenicity* in humans but … *sufficient evidence of carcinogenicity* in experimental animals" (IARC 2013, 30).

Many other common agents appear in this broad category, such as naturally occurring elements in aloe vera, pickled vegetables, coffee, and fruits. Although they are classified in the same group, few of these items receive the same public response and outcry as RF radiation. The phenomenon has been exacerbated by the latest generation of 5G networks and infrastructure.

The lack of official positions from certain trusted organizations, such as the American Cancer Society (ACS) in the United States, may contribute to concern and confusion among skeptical citizens. However, although the ACS has not made an official declaration, it has provided extensive information on its website to highlight the nonionizing nature of 5G, along with some of the nuances of the research process that may have contributed to findings beyond the general consensus (ACS 2022).

The misunderstanding of scientific research and misinterpretation of findings is a significant challenge that policy makers in developing countries must consider. Although consensus exists on the impact of both 5G networks and general cellular usage on health, more research on 5G is needed, particularly for the high band of the spectrum, owing to denser radio deployments (Reardon 2020).

Public concern about 5G infrastructure has been greater than that with previous mobile generations, and this concern was exacerbated by the global COVID-19 pandemic. This issue may seem surprising, as no evidence exists of a causal relationship between 5G and any form of coronavirus transmission. Nonetheless, unproven theories about 5G health risks continue to be spread virally by internet users and through media outlets reporting on the phenomenon (Bruns, Harrington, and Hurcombe 2020).

Guidelines on Radiation Exposure

Over the years, various organizations have developed public health guidelines for radiation exposure to inform policy making on communications technology. WHO formally recognizes two international bodies that have published guidelines on recommendations for overall exposure to RF radiation: the International Commission on Non-Ionizing Radiation Protection (ICNIRP) and the Institute of Electrical and Electronics Engineers (IEEE). Other professional bodies also provide international standards for limiting human exposure to EMFs (refer to appendix A). On the health implications of EMF exposure from 5G, WHO anticipates no adverse consequences for public health from overall exposure that remains below international guidelines (2020) and is expected to publish its study of health risk assessments of RF radiation in 2022 (WHO 2020).

For 5G technologies, ICNIRP (n.d.) notes the following:

> Another general characteristic of RF EMFs is that the higher the frequency, the lower the depth of penetration of the EMFs into the body. As 5G technologies can utilize higher EMF frequencies (>24 GHz) in addition to those currently used (<4 GHz), power from those higher frequencies will be primarily absorbed more superficially than that from previous mobile telecommunications technologies. However, although the proportion of power that is absorbed superficially (as opposed to deeper in the body) is larger for the higher frequencies, the ICNIRP (2020) restrictions have been set to ensure that the resultant peak spatial power will remain far lower than that required to adversely affect health. Accordingly, 5G exposures will not cause any harm providing that they adhere to the ICNIRP (2020) guidelines.

Critics have pointed out that the safety limits of these guidelines have been based on tissue heating and have largely ignored the secondary effects of exposure to RF radiation (Belpomme et al. 2018). Some critics are also concerned that 5G network deployments are taking place without precautions, including a reexamination of scientific studies that confirm the health risks of EMF exposure (Hardell and Carlberg 2020). These collective concerns have led to efforts to halt 5G deployment in some jurisdictions—including the European Union, where 406 scientists and medical doctors signed a 2017 appeal requesting a moratorium on 5G deployments until a full study can be carried out to confirm that 5G does not present health risks to humans (Nyberg and Hardell n.d.). The EU rejected the appeal.

The latest guidelines from ICNIRP, published in March 2020, cover EMF exposures above 6 gigahertz, which includes the mid and high bands of the 5G spectrum. ICNIRP asserts that its guidelines were developed based on thorough reviews of reproducible studies with scientific validity, including the review conducted by WHO in 2014; likewise, the commission claims to have examined all potential impacts of RF radiation on health (ICNIRP 2020).

Managing Health Risks and Perceptions

The lack of clear, consistent, and reassuring messaging from government, industry, academia, and civil society organizations has enabled misinformation and confusion to spread, particularly on the purported danger of EMF exposure from 5G cellular networks. To fight misinformation and dispel confusion, governments and other trusted organizations should mount coordinated messaging campaigns. Governments must manage communications about this topic and refer citizens to trusted institutions that are producing the latest research and guidance on health concerns. Global coordination of funding, research, and messaging among countries would also help increase the clarity and consistency of information.

Setting RF exposure limits to protect public health is an important issue. Policy makers in all countries must assess the trade-offs between strict precautionary standards and higher capital costs for 5G networks. While doing so, they must provide maximum transparency and scientific evidence to inform the public about the potential health implications of RF exposure and the protections offered by regulatory limits. Governments can do this by establishing a working group that is effective in public engagement with the purpose of demystifying, educating, facilitating discussions, and promoting community awareness of past and present research on the safety and environmental impact of nonionizing radiation. The group should also inform the public on the current legal and regulatory frameworks to enforce EMF limits.

Countries should mandate a sufficient level of institutional capacity in knowledge, facilities, staffing, and equipment to monitor and enforce compliance, respond to citizen complaints and inquiries, and investigate cases that might be attributed to nonionizing radiation. Governments should also review the basis of the existing standards for EMF exposure limits and adhere to international guidelines as they develop their national EMF standards. In 2018, the IEEE and the International Electrotechnical Commission formed a joint working group—the International Committee on Electromagnetic Safety—to develop common standards for assessing human exposure. Countries' attention and adherence to the guidelines of the working group, and those of ICNIRP, are strongly recommended.

Regardless of which 5G deployment option a government chooses, policy makers must manage the risks related to cybersecurity, sustainability, and health concerns among the public to capture the full benefits of the new mobile technologies standards and facilitate their deployment. Chapter 4 further examines these risks, referencing the regulatory and policy implications of these risk management efforts.

NOTES

1. Working Group SA3 of the 3rd Generation Partnership Project (3GPP) is responsible for determining 5G network security and privacy requirements and specifying security architectures and protocols. As noted in chapter 1, each generation of mobile technologies' standards is defined by a series of vision statements and performance requirements to which countries agree through the International Telecommunication Union. 3GPP then transforms those agreements into detailed technical standards for purposes of product and service development. SA3 ensures the availability of the cryptographic algorithms included in the specifications and accommodates, as far as practical, regional variations in security objectives and privacy requirements set by national regulators. In parallel, the fraud and security group of the GSM Association proposes updates to existing standards and develops new draft standards specific to the security and anti-fraud aspects of 5G. The GSM Association is the trade association for the mobile communications industry (refer to www.gsma.com).

2. Signaling System 7 was designed in the 1980s, when the number of telecommunication networks in the world was much smaller, being made up of national carriers and large international companies. Trust among network operators was assumed, and interconnection was based on this trust, with no authentication built in. Currently, with many operators, roaming arrangements are still based on trust. Developed in 1998, Diameter is an authentication, authorization, and accounting protocol for electronic networks. The 3GPP uses Diameter in many interfaces, most notably in Long-Term Evolution, as new commands and attributes can be defined for each interface. Diameter includes no encryption but can be protected by transport-level security (Internet Protocol Security or Transport Layer Security).

3. For more details on how 5G can be secured through open RAN architecture, refer to https://www.cisa.gov/sites/default/files/publications/open-radio-access-network-security-considerations_508.pdf.

4. 5G threat surface analysis is a complex and laborious exercise conducted in just a few developed countries. This section offers a general understanding of the topic. For a more detailed analysis, refer to ENISA (2019).

5. According to Huber (2019), "security can be patchy for some IoT devices, especially low-cost and low-powered items. Hackers can use technology to scan hundreds of thousands of devices for weak security, such as those with the default passwords—'admin,' 'guest,' or 'password'— that they were sold with. […] The likelihood of finding an IoT device that hasn't been set up properly, or with a weak password, is quite high."

6. A *botnet* is a collection of internet-connected devices infected with malware that allows hackers to control them. Cyber criminals use botnets to instigate attacks whose purpose may be to cause credential leaks, unauthorized access, data theft, or service denial. For more details, refer to Akamai (2017). A denial-of-service attack occurs when legitimate users are unable to access information systems, devices, or other network resources due to the actions of a malicious cyber threat actor. The affected services may include email, websites, online accounts (for example, banking), or other services that rely on the affected computer or network. Denial of service is accomplished by flooding the targeted host or network with so much traffic that the target cannot respond or simply crashes, preventing access for legitimate users. For more details, refer to CISA (2019b).

7. Roaming agreements allow operators to control access for roaming subscribers and manage roaming services. The home operator entrusts the roaming operator with serving its subscribers, and interconnection is used to support that service.

8. The vulnerability allows attackers to impersonate a mobile device, enabling them to register for services in the owner's name.

9. The Science Based Targets initiative "drives ambitious climate action in the private sector by enabling companies to set science-based emissions reduction targets"; refer to https://sciencebasedtargets.org. General European Strategic Investments is a public-private partnership of ICT companies and international organizations focused on steering the ICT sector toward sustainable development; refer to http://gesi.org.

10. For more information on the Basel Convention on the Control of Transboundary Movements of Hazardous Wastes and Their Disposal, refer to http://www.basel.int/. The 2019 technical guidelines on e-waste are available at http://www.basel.int/Implementation/TechnicalMatters/DevelopmentofTechnicalGuidelines/TechnicalGuidelines/tabid/8025/Default.aspx.

BIBLIOGRAPHY

3GPP (3rd Generation Partnership Project). 2019. "Study on the Support of 256-Bit Algorithms for 5G." *Technical Report 33.841*. 3GPP, Sophia Antipolis Technology Park, France. https://portal.3gpp.org/desktopmodules/Specifications /SpecificationDetails.aspx?specificationId=3422.

3GPP (3rd Generation Partnership Project). 2020a. "3GPP System Architecture Evolution (SAE); Security Architecture." *Technical Specification 33.401*. 3GPP, Sophia Antipolis Technology Park, France. https://portal.3gpp.org/desktopmodules /Specifications/SpecificationDetails.aspx?specificationId=2296.

3GPP (3rd Generation Partnership Project). 2020b. "Security Architecture and Procedures for 5G System." *Technical Specification 33.501*. 3GPP, Sophia Antipolis Technology Park, France. https://portal.3gpp.org/desktopmodules/Specifications /SpecificationDetails.aspx?specificationId=3169.

ACS (American Cancer Society). 2022. "Cellular (Cell) Phones." March 31. https:// www.cancer.org/cancer/risk-prevention/radiation-exposure/cellular-phones .html#additional_resources.

Akamai. 2017. "What Is a Botnet Attack?" Akamai Technologies, Cambridge, MA. https://www.akamai.com/us/en/resources/what-is-a-botnet.jsp.

Amin, Rami, Bertram Boie, Natalija Gelvanovska-Garcia, and Sandra V. Sargent. 2020. "How COVID-19 Has Exposed Cyber Risks in the Health Sector: Why a Paradigm Shift Is Needed for Building Cybersecurity Resilience." *Analytical Insights 1*, World Bank, Washington, DC. http://pubdocs.worldbank.org/en/989431596041514122 /July-2020-Digital-Development-Thought-Leadership.pdf.

Belpomme, Dominique, Lennart Hardell, Igor Belyaev, Ernesto Burgio, and David O. Carpenter. 2018. "Thermal and Non-Thermal Health Effects of Low Intensity Non-Ionizing Radiation: An International Perspective." *Environmental Pollution* 242: 643–58.

Borgaonkar, Ravishankar, Lucca Hirschi, Shinjo Park, and Altaf Shaik. 2019. "New Privacy Threat on 3G, 4G and Upcoming 5G AKA Protocols." *Proceedings on Privacy Enhancing Technologies*, 2019 (3), 108–27. https://doi.org//10.2478 /popets-2019-0039.

Bruns, Axel, Stephen Harrington, and Edward Hurcombe. 2020. "Corona? 5G? or Both? The Dynamics of COVID-19/5G Conspiracy Theories on Facebook." *Media International Australia* 177 (1): 12–29. https://doi.org/10.1177/1329878X20946113.

CISA (Cybersecurity and Infrastructure Security Agency). 2019a. *Security Tip ST04-015: Understanding Denial-of-Service Attacks*. Arlington, VA: CISA. https://www.us-cert .gov/ncas/tips/ST04-015.

CISA (Cybersecurity and Infrastructure Security Agency). 2019b. *Overview of Risks Introduced by 5G Adoption in the United States*. Arlington, VA: CISA. https://www .cisa.gov/sites/default/files/publications/19_0731_cisa_5th-generation-mobile -networks-overview_0.pdf.

Dalul, Suzana. 2020. "To Solve the Smartphone E-Waste Problem, We First Need Fewer Disposable Devices." *Android Authority*, July 26. https://www.androidauthority .com/e-waste-smartphones-1133322/.

Engel, Tobias. 2014. "SS7. Locate. Track. Manipulate." *Chaos Computer Club*. https:// berlin.ccc.de/~tobias/31c3-ss7-locate-track-manipulate.pdf.

ENISA (European Union Agency for Cybersecurity). 2019. *Threat Landscape for 5G Networks*. https://www.enisa.europa.eu/publications/enisa-threat-landscape-for -5g-networks.

ETSI (European Telecommunications Standards Institute). 2015. *Quantum Safe Cryptography and Security*. ETSI White Paper No. 8. Sophia Antipolis Technology Park, France: ETSI. https://www.etsi.org/images/files/ETSIWhitePapers /QuantumSafeWhitepaper.pdf.

Fixler, Dror, and Adam Weinberg. 2020. "5G Threats Analyis." Background paper prepared for *This 5G Flagship Report,* World Bank, Washington, DC.

Gartner. 2019. "Gartner Predicts Outdoor Surveillance Cameras Will Be Largest Market for 5G Internet of Things Solutions Over Next Three Years." https://www.gartner.com/en/newsroom/press-releases/2019-10-17-gartner-predicts-outdoor-surveillance-cameras-will-be.

Grad, Peter. 2020. "Report Linking 5G to COVID-19 Swiftly Debunked." *Science X Network,* July 27. https://medicalxpress.com/news/2020-07-linking-5g-covid-swiftly-debunked.html.

GSM Association. 2019. *The 5G Guide: A Reference for Operators.* London: GSM Association. https://www.gsma.com/wp-content/uploads/2019/04/The-5G-Guide_GSMA_2019_04_29_compressed.pdf.

GSM Association. 2022. *Mobile Net Zero: State of the Industry on Climate Action 2022.* London: GSM Association. https://www.gsma.com/betterfuture/wp-content/uploads/2022/05/Moble-Net-Zero-State-of-the-Industry-on-Climate-Action-2022.pdf.

Gye, Mung Chan, and Chan Jin Park. 2012. "Effect of Electromagnetic Field Exposure on the Reproductive System." *Clinical and Experimental Reproductive Medicine* 39 (1): 1–9. https://www.ncbi.nlm.nih.gov/pmc/articles/PMC3341445/.

Hardell, Lennart, and Michael Carlberg. 2020. "Health Risks from Radiofrequency Radiation, Including 5G, Should Be Assessed by Experts with No Conflicts of Interest." *Oncology Letters* 20 (4): 1.

Huber, Nick. 2019. "A Hacker's Paradise? 5G and Cyber Security." *Financial Times.* October 13. https://www.ft.com/content/74edc076-ca6f-11e9-af46-b09e8bfe60c0.

Hussain, Syed Rafiul, Mitziu Echeverria, Omar Chowdhury, Ninghui Li, and Elisa Bertino. 2019. "Privacy Attacks to the 4G and 5G Cellular Paging Protocols Using Side Channel Information." Presented at the Network and Distributed System Security Symposium. https://www.ndss-symposium.org/wp-content/uploads/2019/02/ndss2019_05B-5_Hussain_paper.pdf.

IARC (International Agency for Research on Cancer). 2013. "IARC Monographs on the Evaluation of Carcinogenic Risks to Humans." In *Volume 102. Non-Ionizing Radiation, Part 2: Radiofrequency Electromagnetic Fields.* https://publications.iarc.fr/126.

ICNIRP (International Commission on Non-Ionizing Radiation Protection). 2020. "FAQ." ICNIRP, Oberschleissheim, Germany. https://www.icnirp.org/en/rf-faq/index.html#:~:text=Does%20ICNIRP%20consider%20non%2Dthermal,the%20exposure%20and%20the%20body.

ICNIRP (International Commission on Non-Ionizing Radiation Protection). n.d. "5G Radiofrequency—RF EMF." ICNIRP, Oberschleissheim, Germany. https://www.icnirp.org/en/applications/5g/5g.html.

IPCC (Intergovernmental Panel on Climate Change). 2018. *Global Warming of 1.5°C. An IPCC Special Report on the Impacts of Global Warming of 1.5°C above Pre-industrial Levels and Related Global Greenhouse Gas Emission Pathways, in the Context of Strengthening the Global Response to the Threat of Climate Change, Sustainable Development, and Efforts to Eradicate Poverty.* Masson-Delmotte, Valérie, Panmao Zhai, Hans-Otto Pörtner, Debra C. Roberts, James Skea, Priyadarshi R. Shukla, Anna Pirani, et al. (eds.). Cambridge, UK: Cambridge University Press, https://doi.org/10.1017/9781009157940.

ITU-T (International Telecommunication Union Telecommunications Standardization Sector). 2020. "L.1470: Greenhouse Gas Emissions Trajectories for the Information and Communication Technology Sector Compatible with the UNFCCC Paris Agreement." https://www.itu.int/ITU-T/recommendations/rec.aspx?rec=14084.

Knight, Ben. 2020. "5G: The True Cost Will Be Measured in E-Waste." *UNSW Newsroom.* August 4. https://newsroom.unsw.edu.au/news/art-architecture-design/5g-true-cost-will-be-measured-e-waste.

Kuchler, Hannah. 2017. "Three Plead Guilty to Causing Massive US Cyberattack." *Financial Times.* December 13. https://www.ft.com/content/3b9ff338-e05c-11e7 -8f9f-de1c2175f5ce.

Masoudi, Meysam, Mohammad Galal Khafagy, Alberto Conte, Ali El-Amine, Brian Françoise, Chayan Nadjahi, et al. 2019. "Green Mobile Networks for 5G and Beyond." *IEEE Access,* 107270–107299. https://ieeexplore.ieee.org/document /8786138.

Nakarmi, Prajwol Kuman, and Karl Norrman. 2018. "Detecting False Base Stations in Mobile Networks." *Ericsson Blog*, June 15. https://www.ericsson.com/en/blog /2018/6/detecting-false-base-stations-in-mobile-networks.

Nigeria: NCC Clears Doubts Over 5G, COVID-19 and Security. 2020 (April 6). https://link.gale.com/apps/doc/A619562936/AONE?u=anon~f70c74f1&sid =sitemap&xid=9ca2b00e.

NIS Cooperation Group. 2019. *EU Coordinated Risk Assessment of the Cybersecurity of 5G Networks.* Brussels, Belgium: European Commission. https://ec.europa.eu /newsroom/dae/document.cfm?doc_id=62132.

NIST (National Institute of Standards and Technology). 2017. *Guide to LTE Security.* Special Publication 800-187. Gaithersburg, MD: NIST. https://www.nist.gov /publications/guide-lte-security.

Nyberg, Rainer, and Lennart Hardell. n.d. "5G Appeal." http://www.5gappeal.eu /about/.

Peisa, Janne, Patrik Persson, Stefan Parkvall, Erik Dahlman, Asbjørn Grøvlen, Christian Hoymann, and Dirk Gerstenberger. 2020. "5G Evolution: 3GPP Releases 16 & 17 Overview." *Ericsson Technology Review.* March 9. https://www.ericsson.com/en /reports-and-papers/ericsson-technology-review/articles/5g-nr-evolution.

Positive Technologies. 2019. "5G Signaling Networks: Blast from the Past." December 18. https://positive-tech.com/research/5g-signaling-networks/.

Reardon, Marguerite. 2020. "Is 5G Making You Sick? Probably Not." *CNET*, July 30. https://www.cnet.com/news/is-5g-making-you-sick-probably-not/.

Reichert, Corinne. 2020. "5G Coronavirus Conspiracy Theory Leads to 77 Mobile Towers Burned in UK, Report Says." *CNET*, May 7. https://www.cnet.com/health/5g -coronavirus-conspiracy-theory-sees-77-mobile-towers-burned-report-says/.

Rupprecht, David, Katharina Kohls, Thorsten Holz, and Christina Popper. 2020. "IMP4GT: IMPersonation Attacks in 4G NeTworks." *Network and Distributed Systems Security Symposium*, February 23–26, San Diego, CA. https://www.ndss-symposium .org/wp-content/uploads/2020/02/24283.pdf.

UK Office of Communication. n.d. "5G Mobile Technology: A Guide" https://www .ofcom.org.uk/__data/assets/pdf_file/0015/202065/5g-guide.pdf.

US Centers for Disease Control and Prevention. 2014. "Frequently Asked Questions about Cell Phones and Your Health." June 9. US Centers for Disease Control and Prevention, Washington, DC. https://www.cdc.gov/nceh/radiation/cell_phones ._FAQ.html.

US Environmental Protection Agency. n.d. "Cleaning up Electronic Waste (E-Waste)." US Federal Communications Commission, Washington, DC. https://www.epa.gov /international-cooperation/cleaning-electronic-waste-e-waste.

US Federal Communications Commission. 2020. "Wireless Devices and Health Concerns." November 4. US Federal Communications Commission, Washington, DC. https://www.fcc.gov/consumers/guides/wireless-devices-and-health-concerns.

US Food and Drug Administration. 2022. *Do Cell Phones Pose a Health Hazard?* Washington, DC: US Food and Drug Administration. https://www.fda.gov/radiation -emitting-products/cell-phones/do-cell-phones-pose-health-hazard.

Waterson, Jim, and Alex Hern. 2020. "How False Claims about 5G Health Risks Spread into the Mainstream." *The Guardian,* April 7. https://www.theguardian .com/technology/2020/apr/07/how-false-claims-about-5g-health-risks-spread -into-the-mainstream.

Wheeler, Tom, and David Simpson. 2019. *Why 5G Requires New Approaches to Cybersecurity.* Washington, DC: Brookings Institution. https://www.brookings.edu /research/why-5g-requires-new-approaches-to-cybersecurity/.

WHO (World Health Organization). 2020. *5G Mobile Networks and Health.* Geneva: WHO. https://www.who.int/westernpacific/news/q-a-detail/5g-mobile-networks -and-health.

WHO (World Health Organization). n.d. *Coronavirus Disease (COVID-19) Advice for the Public: Mythbusters.* Geneva: WHO. https://www.who.int/emergencies/diseases /novel-coronavirus-2019/advice-for-public/myth-busters#5g.

Regulatory Imperatives, Policy Challenges, and Recommendations for Action

KEY MESSAGES

- Decisions about policy, regulation, and investment all have major implications for how the fifth-generation (5G) mobile networks are built, operated, and integrated into daily life.

- Spectrum is a finite resource that must be allocated wisely and deliberately as advanced mobile networks develop and take shape, particularly with new business models and industry players in sectors beyond telecommunications. It is essential for all types of mobile communications and many other scientific, military, civil, and commercial activities.

- Regulators have an important role in facilitating the deployment and operation of 5G networks, in developing cell sites and fiber networks, in sharing infrastructure and spectrum, and in influencing the pace of innovation.

INTRODUCTION

Mobile telecommunications operators have been the primary drivers of the expansion and uptake of previous generations of mobile technology. The same will be true for 5G, with operators worldwide expected to invest close to US$1 trillion in 5G networks from 2020 to 2025 (GSM Association 2020). Additional investments by other industries are expected as well. Most investment will occur in high-income countries, but developing countries will benefit, too (refer to figure 4.1). Yet, investment is only one factor in the broad array of decisions about policy and regulation that will shape how

FIGURE 4.1 Capital spending for 5G, 2020–25 (US$, billions)

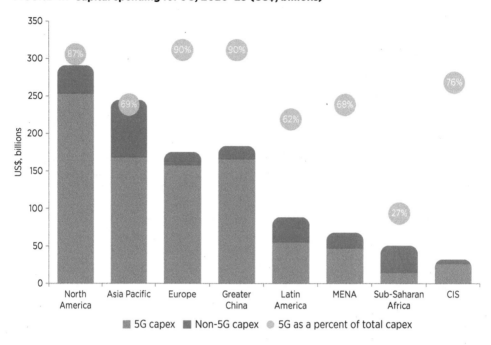

Source: GSM Association 2020.
Note: capex = capital expenditures; CIS = Commonwealth of Independent States; MENA = Middle East and North Africa.

5G networks are built, operated, and integrated into personal life, industry, trade, and government.

Spectrum comes first. Governments are responsible for deciding how spectrum is allocated and used. Decisions about how it is assigned to licensees have been, and will continue to be, a powerful driver of the evolution of mobile technologies. Access to sites on which to build 5G network equipment is another key determinant of both the technical and financial performance of networks. Governments can play a major part through policy and regulation, but also—in many countries—as providers of backhaul infrastructure. Finally, governments must gain public support for public investments; deployment requirements; and the health, safety, and security standards of new telecommunications infrastructure, as discussed in chapter 3.

Key policy decisions have important implications for how far and how quickly 5G network coverage will spread. Many of these considerations are not specific to 5G, of course, because they applied equally to earlier generations of the technology and continue to be relevant. However, 5G creates new policy questions for governments, such as the following:

• The high-frequency bands allocated to 5G will require many more radio sites than the lower-frequency bands. For this requirement to be financially and operationally-feasible, it may be necessary to provide widespread access to public infrastructure in ways not seen in previous mobile generations. If extending mobile coverage is an important policy goal, then a blend of fourth-generation (4G) mobile network technologies and 5G may be needed.

- The potential for vertical applications and uses of 5G technology in various sectors means that new ways of assigning and licensing spectrum may be required.

- It is possible that new players will emerge on their own or through strategic partnerships. This could require regulators to reconsider policies related to market structure and competition.

- New or heightened areas of risk, such as cybersecurity, arise from the increased amount of data carried by 5G and the depth of the technology's integration into supply chains (for example, for smart applications in city management and the food chain).

These new aspects of 5G create significant uncertainty about how the technology will develop and what its economic implications are likely to be, especially in developing countries.

The policy and regulatory decisions that governments make for 5G are likely to reverberate for many years to come. For example, it may be difficult to change spectrum assignments once they are made (although often a standard clause in licenses exists that enables policy makers to make changes in the public interest). Furthermore, once they are established, it is difficult to shift mobile market structures that develop in response to spectrum and regulatory decisions. These challenges will grow as 5G networks penetrate sectors such as manufacturing, agriculture, or extractive industries, where they have the potential to yield major benefits.

When identifying policy issues and making recommendations at this early stage of 5G deployment, it is important to remember both the uncertainty and the long-term consequences of decisions that are made now. The priority should be to focus on policies that can be described as "no-regret" decisions. These decisions would be beneficial for a country to take across a broad range of network rollout scenarios. In many cases, they will be beneficial for mobile network deployment and service delivery in general, while also advancing 5G.

Therefore, the twofold objective of this chapter is, first, to highlight policy and regulatory challenges governments are likely to face as they seek to facilitate development of the 5G mobile network, and second, to suggest a strategy toward 5G, including a set of no-regret steps that governments can take in each domain regardless of how far they may be from widescale 5G deployment.

SPECTRUM MANAGEMENT

Spectrum is essential for all types of mobile communications, as well as other uses, such as scientific research, military applications, broadcasting, civil aviation, search and rescue operations, and delivery services, among others. An international spectrum allocation process administered by the International Telecommunication Union determines which bands are to be used for which purposes. Spectrum frequencies have different technical properties, but competing uses for spectrum bands exist, and national governments should make the necessary trade-offs between those uses.

In many cases, the frequency bands with optimal technical characteristics (for example, those with good propagation, low interference, and in-building penetration) have already been assigned to other uses and users. National governments and regulators must decide how to deal with new applications such as 5G and how to balance the interests of existing and new users. In some cases, assignments of spectrum to 5G may be constrained. In other cases, the existing users may be moved to other frequencies to free up the optimal spectrum for 5G.

Challenges and Considerations

This process is challenging for regulators and requires careful analysis and a mix of approaches, including the following:

- Replacing analog television with digital so that large parts of the ultra-high frequency band can be reallocated to cellular technologies such as 4G and 5G.

- Allowing mobile network operators (MNOs) to repurpose existing second-generation (2G) mobile network technologies, third-generation (3G) mobile network technologies, and 4G channels for 5G by making licenses technology neutral and permitting dynamic spectrum sharing.

- Moving satellite links out of the 3.4–4.0 gigahertz (GHz) band so that parts of it can become a 5G "pioneer" band.

- Reallocating incumbents out of large parts of the 24–28 GHz millimeter wave (mmWave) band so it can be used for 5G.

Except for some parts of the mmWave, all these changes have already required moving incumbent licensees to different bands in some countries. Because most of the mid- and low-band incumbents still have valid licenses, these operators must be compensated for the loss of their channels and for having to buy new equipment. This process is costly, disruptive, and time-consuming, not only due to technical reasons but also because it requires negotiations (clearing the 3.5 GHz and 700 megahertz [MHz] bands took several years). Long-range planning is essential everywhere for responsive and efficient spectrum management.

The challenge of deciding between different uses for spectrum has been faced by regulators with earlier generations of mobile technology. 5G presents specific challenges but also creates new possibilities for regulators:

- The amount of spectrum needed for 5G to fulfill its full performance potential is still not completely clear. However, the mobile industry has proposed that each MNO should receive 80–100 MHz of contiguous bandwidth in the mid band (1.5–7.12 GHz) and 1 GHz in the high band (24–40 GHz) to reach optimal 5G network performance (GSM Association 2017). These requirements are more than the total bandwidth granted for all previous generations of international mobile telecommunications, raising questions about the extent to which spectrum frequencies should be dedicated to this one use at the expense of others. In practice, 27 countries have assigned contiguous bandwidth of less than 100 MHz per operator in the mid band, and 18 countries have assigned less than 1 GHz of contiguous bandwidth to a single operator in the high band.[1]

- The technical specifications of 5G imply other viable forms of spectrum management and licensing. 5G can aggregate licensed and unlicensed bandwidth, allowing the creation of virtual channels for more throughput.

- In addition, in developed countries, a trend exists toward "light licensing"— also referred to as a "third way"—which combines attributes of licensing and license exemption. Finding the right balance among licensed, light licensed, unlicensed, and aggregated spectrum is a complicating factor for regulators, although having more options can provide needed flexibility.[2]

The deployment of 5G in sectors other than telecommunications (particularly in vertically structured industries) further complicates spectrum management. However, it also yields new opportunities to support innovative business models and new use cases for spectrum in other sectors and industries beyond telecommunications. To accommodate 5G—and thereby unlock its benefits in those sectors—governments must design policies to facilitate access to specific frequencies for use by non-MNO users. The potential benefits are great as new use cases continue to develop over time (refer to chapter 2) but can have broad implications for spectrum management. For example, a spectrum set-aside policy could reduce the amount of contiguous spectrum available for other users, which could reduce performance or raise costs.

Subnational, regional, and local operators are unlikely to seek national licenses but would look for licenses covering a specific geographical area. Designing geographically defined licenses for subnational coverage would be a significant departure from previous practice for many countries, and so far, countries are taking various approaches. At least 11 countries (including Australia, Finland, and Japan) have mechanisms in place to grant spectrum for industrial 5G networks, mostly through local licenses using a first-come, first-served approach for selected bands. In some countries, licensees have obligations to lease spectrum for private networks (for example, Denmark for 3,740–3,800 MHz) or provide services to verticals (for example, France for 3.4–3.8 GHz), as noted in table 4.1. In Germany, a use-or-lose provision applies to the 24.25–27.5 GHz band, and in the United Kingdom, local shared access is allowed in certain bands (for example, 3.8–4.2 GHz, 1,781.7–1,785/1,876.7–1,880 MHz, and 2,390–2,400 MHz).[3]

TABLE 4.1 **Authorization regimes for private 5G networks**

Authorization regime	Description
Licensed spectrum	National regulator assigns spectrum for local private networks. *Examples:* France[a] and Germany.[b]
Subleased spectrum	Public operator provides a private network as a service using the spectrum it has been assigned. *Example:* Nippon Telegraph and Telephone Corporation in Japan.[c] Public operator subleases the spectrum it has been assigned to the private network operator. *Example:* Denmark, 3,740–3,800 MHz.[d]
Shared spectrum	More than one organization uses the same spectrum range within a given geographical area. *Examples:* 3.5 GHz band in the United States[e] and 3.8–4.2 GHz, 1781.7–1,785/1,876.7–1,880 MHz, and 2,390–2,400 MHz bands in the United Kingdom.[f]
Unlicensed spectrum	Organizations can use spectrum that is freely available for public use to enable use of LTE or 5G, such as 2.4 GHz and 5 GHz widely used for Wi-Fi. *Example:* 6 GHz band in the United States.[g]

Sources: Cullen International 2021; typology adapted from GSM Association 2020.
Note: 5G = fifth generation mobile network technologies; GHz = gigahertz; LTE = Long-Term Evolution; MHz = megahertz.
a. For France's allocation details for the 3.4–3.8 GHz band, refer to https://en.arcep.fr/news/press-releases/view/n/5g -20.html.
b. For Germany's provision details, refer to https://www.bundesnetzagentur.de/DE/Sachgebiete/Telekommunikation /Unternehmen_Institutionen/Frequenzen/OeffentlicheNetze/LokaleNetze/lokalenetze-node.html.
c. Nippon Telegraph and Telephone Corporation launched a private 5G "network-as-a-service" platform in August 2021. Refer to https://hello.global.ntt/en-us/newsroom/ntt-launches-first-globally-available-private-5g-network-as-a-service -platform.
d. For standard contract details for the 3,740–3,800 MHz band in Denmark, refer to https://ens.dk/sites/ens.dk/files /Tele/annex_m_-_standard_contract_for_leasing_spectrum.pdf.
e. For mid band 5G sharing in the United States, refer to https://www.fcc.gov/document/fcc-opens-100-megahertz -mid-band-spectrum-5g-0.
f. For shared spectrum licensing in the United Kingdom, refer to https://www.ofcom.org.uk/manage-your-licence /radiocommunication-licences/shared-access.
g. Announced by the US Federal Communications Commission in April 2020; refer to https://www.fcc.gov/document /fcc-opens-6-ghz-band-wi-fi-and-other-unlicensed-uses-0.

Another consideration is that the range of industrial uses is likely to evolve over time. Therefore, it would not be prudent to issue all available spectrum to current users, as doing so might dampen innovation and alternative uses in the future.

If governments enable spectrum access for non-MNOs, they also should design a process for selecting these spectrum users and drafting the rules that will apply to them. It may also be necessary to decide whether alternative licensing models should be applied using emerging approaches such as sharing spectrum with other users, reserving databases, and managing dynamic interference.

Other ways exist for non-MNO users (for example, manufacturers) to gain exclusive access to portions of spectrum or network capacity. Network slicing allows users to have full access to specific segments of network capacity provided by operators, as noted in chapter 1. This process has the advantage that private users do not have to build and operate their own network. In addition, the regulator does not have to design special spectrum management arrangements for private users. However, as of 2023, it is unclear how slicing would work in practice and whether suitable commercial offers will appear.

In designing spectrum strategies, regulators should be aware of the different models of spectrum usage, as shown in table 4.1 and incorporate them into their planning and consultation.

The issue of assigning 5G spectrum to entities other than MNOs is not limited to industrial uses. Critical public services, such as ambulances, fire departments, and electric utilities, could also be allocated specific bands and frequency licenses. In most countries, utilities, public safety, and emergency services already have exclusive or shared national allocations in which non–*3rd Generation Partnership Project* network technologies are deployed (TETRA, P25, MPT-1327, and others). In principle, 5G could be implemented within those existing allocations (if the technical standards encompass those frequency ranges and manufacturers provide equipment), but regulators would have to modify current channeling schemes to accommodate 5G's wider channels. Compatibility issues also may exist between 5G and legacy networks.

Many new areas of spectrum usage are being considered for specialized public services. For example, intelligent transport systems and the internet of things networks for monitoring urban environments would be compelling candidates for 5G spectrum, and regulators might consider assigning dedicated frequency bands for their exploitation.

Alternatively, it may be possible for such specialized public uses of 5G spectrum to be obtained more efficiently through spectrum sharing or network slicing than through the exclusive licensing of spectrum. Regulators should retain flexibility on this and observe how the technology and business models evolve over time.

Recommendations for Spectrum Management

Recommendations for spectrum management include expanding institutional capacity; establishing a national spectrum roadmap; ensuring the timely availability of 5G spectrum; determining the appropriate spectrum assignment method and pricing; licensing spectrum wisely; and promoting innovation in spectrum-sharing models, including spectrum subleasing.

EXPAND INSTITUTIONAL CAPACITY FOR SPECTRUM MANAGEMENT

Governments should improve their capacity to manage spectrum so they can exploit 5G's promise and meet the challenges it poses. Better management will bring broad benefits to the telecommunications sector and beyond.

Broadly, spectrum management can be divided into four parts: planning, engineering, authorization, and monitoring and compliance (Blackman and Srivastava 2011). 5G presents challenges in each of these areas. The frequency bands that are appropriate for use in 5G require technical analysis and a process for clearing bands that have existing users. New models of spectrum licensing and sharing are also important. Spectrum authorities will require new areas of expertise in technical analysis, legal work, and licensing for successful implementation.

The potential for 5G to bring new spectrum users into the market will also create challenges for the regulatory authorities. First, the design of assignment processes that offer opportunities for new players without undue distortions of the existing market will require careful consideration and consultation. Second, once spectrum has been assigned to these users, enhanced capacity will be required to monitor how it is being used, prevent interference with other users, and take enforcement actions where necessary.

Transparency and accountability in how spectrum is planned, assigned, and monitored has always been important, but the deployment of 5G increases the need for both. If, as is likely, 5G permits further integration of mobile networks into commercial activity and public service delivery, more transparency, accountability, and stability in the way spectrum is managed will be needed.

Establish a National Spectrum Roadmap

Predictability and transparency in spectrum planning have always been beneficial for licensees and, ultimately, for customers. However, 5G makes them even more important than they were in earlier generations of mobile technology. Although 5G has begun to deploy, it is still in the early stages in many developing countries, and considerable uncertainty remains about how it will evolve and whether the novel spectrum arrangements discussed earlier will develop. Forward planning, transparency, and consultation with the private sector will be more important than ever with 5G.

Regulators should establish a forward-looking, flexible, and sustainable roadmap to inform stakeholders of a country's strategic direction of spectrum management. The roadmap should be detailed for a few years into the future, with indications of changes in allocation that are being discussed for the next 2–5 years, considering the long lead times for opening new bands and the challenges of broader global harmonization.[4] National spectrum roadmaps should include all spectrum users, including governments, industries, businesses, and individuals. The issues to be clarified should incorporate whether to license non-MNOs to use international mobile telecommunications spectrum, how to regulate infrastructure sharing, and whether to permit spectrum subleases. Roadmaps should also encompass plans for license exempt spectrum and innovative approaches for spectrum sharing.

Finally, roadmaps should specify what spectrum will become available and when, as well as other relevant details such as the license duration, its transferability, whether the licensed frequency is shared or exclusive, and deployment requirements. The geographical scope of spectrum licenses could also be considered in roadmaps, particularly in instances where it may be easier to move away from the general practice of national spectrum licenses toward more geographically targeted licenses (with differentiated pricing to lower costs in rural or nonviable areas).

The complexity of spectrum planning, the continually evolving business and technical factors, and the need for trade-offs between different users mean that spectrum roadmaps should be developed in close consultation with stakeholders, especially with industry and the private sector. Input from current and future users will enable the regulatory authorities to make balanced decisions about future spectrum planning. For example, the Republic of Korea's 2017 K-ICT Spectrum Plan and the United Kingdom's regularly published Spectrum Management Strategy statements[5] can demonstrate how governments can consider future 5G networks within their overall spectrum plans for the short- to medium-term time horizons.

Ensure Timely Availability of 5G Spectrum

Global agreements on spectrum allocation and technical specifications continue to evolve. To the extent possible, developing countries should participate in this dialogue to ensure that their interests are represented.

It is in the national interest to ensure that, once decisions are made on spectrum to be allocated for 5G, the allocated spectrum should be made available as soon as possible. This issue may involve clearing existing users from the bands, which is often a difficult and costly exercise. The process is helped by clear and transparent planning and communication by the spectrum authority, in consultation with all users. Another essential element is internal agreement within the government on the plans for shifting existing users of spectrum bands, which allows for effective decision-making and faster implementation.

The structure of spectrum bands assigned to individual users of 5G is a critical decision to be made by spectrum authorities. Although it is desirable to ensure that the spectrum bands assigned to 5G users are sufficiently wide to allow the technology to deliver its full potential, a decision to make large assignments of spectrum to individual licensees may limit the number of viable 5G players in the market. This issue has consequences for competition and, ultimately, for end users. These decisions should be weighed carefully.

Determine the Appropriate Spectrum Assignment Method and Pricing

Choosing the right mechanism for assigning 5G spectrum to individual licensees is critical. Governments should explore which of the various methods works best, considering their country's circumstances. If an auction is the preferred method, the regulatory authorities should pay careful attention to the design to minimize market distortions and ensure that the auction delivers policy objectives.

Some features of auction design merit detailed consideration. For example, reserve prices are often set with the intention of maximizing auction revenue. However, excessively high reserve prices can have the opposite effect, by discouraging participation in the auction and thus reducing competition. Caps and floors also require careful analysis. These are designed to ensure that the outcome of the auction does not create an unbalanced market, which is a sensible objective. However, how the design of caps and floors translates into the overall spectrum endowments of participants can be complex and requires detailed consideration and consultation.

The eligibility rules that determine what types of entities can bid for specific bands of spectrum are another aspect of auction design that affects how 5G will develop. Regulatory authorities can use these rules to achieve specific outcomes, such as attracting new entrants into the market. However, the eligibility rules can also have the effect of suppressing competition in the auction for certain bands of spectrum. Achieving the right balance between ensuring sustainable competition between existing players and encouraging new entries into the market is a challenge for spectrum authorities around the world, and in the 5G era, it will be more important than ever to consider how new use cases, non-telecommunications sectors, and business models are integrating 5G.

License Spectrum Wisely

Spectrum licenses typically contain a wide range of conditions pertaining to how the spectrum can be used, including, among others, coverage obligations; quality of service requirements; rights, restrictions, and requirements on sharing; and measures to prevent interference. These conditions are a key tool through which regulatory authorities can achieve policy objectives.

Coverage obligations are used by most regulators around the world. These require the licensee to provide services in specified areas or over a certain percentage of a country. They can effectively and transparently ensure that parts of a country that may not be commercially attractive are served by mobile operators. Broadly, the greater the coverage obligations, the lower the price that a bidder is willing to pay for the license. For that reason, regulatory authorities must make a trade-off between achieving wide network coverage and raising revenue for the government.

With 5G, as in earlier generations of mobile networks at the time of their launch, considerable uncertainty surrounds the details of the business case. In developing countries, cases may exist where 5G is not the least expensive way to deliver broadband in marginal areas. In this situation, high coverage targets in 5G licenses could be counterproductive because they would result in limited competition and low prices in auctions. In addition, licensees could fail to meet the coverage targets, resulting in disputes with the regulatory authority. Therefore, regulatory authorities should use caution when designing licenses to ensure that coverage targets are sufficient to meet policy objectives for the sector without undermining the underlying business case.

Another key feature of spectrum licenses is the technology and operational conditions they include. Spectrum authorities have often issued spectrum licenses that include constraints and limitations on the uses of the spectrum, such as which generation of network technology can be deployed on specified frequencies. A better approach is for spectrum licenses to include technical conditions pertaining to power and interference but not to constrain the user to one or more specific technologies—in other words, technology neutral spectrum licensing. In the case of 5G, several spectrum bands can be used for different generations of mobile network technologies. It is more efficient for licensees (usually MNOs) to decide how their spectrum is deployed and define their own upgrade paths.

Promote Innovation in Spectrum-Sharing Models,
Including Spectrum Subleasing

5G creates new opportunities for innovation in sharing spectrum (and other network assets). Sharing can be done in various ways. Some models are based on arrangements made independently by spectrum licensees. Others require more direct involvement by the spectrum authority. In all cases, the spectrum authority has an important role to play in ensuring that sharing works effectively from a technical perspective without lessening competition in the market.

5G has created new demands on spectrum resources that are likely to increase over time as technology develops and the underlying demand for data communications grows. This issue means that spectrum must be used more efficiently than in the past, allowing room for innovation in technology and business models. Interest among stakeholders in spectrum-sharing arrangements is likely to be sustained. Examples of spectrum-sharing rules include "use-or-share" clauses,[6] systems for dynamic spectrum sharing,[7] and frameworks to facilitate sharing in local areas.[8] Regulatory authorities in developing countries should consider these issues in detail as they proceed with 5G spectrum licensing.

REGULATORY FRAMEWORKS

Regulators have an important role to play in supporting the deployment and operation of 5G networks. In many cases, this role is like the one played in earlier generations of networks. In other respects, however, the regulator's role is greater, particularly if 5G is to achieve its full potential for network performance and efficiency and if the new issues raised are to be handled properly.

Cell sites are a major issue for network operators, as they represent most of the total capital costs of deploying a network. The locations and technical specifications of the sites also have a major impact on the performance of the network, affecting a range of technical issues such as broadband speeds, quality of service, and in-building penetration.

The deployment of 5G will have significant implications for operators' cell sites. Operators will need to install 5G radio equipment and, in many cases, upgrade or replace associated equipment such as power supplies, temperature control systems, and other components. The higher frequency spectrum that 5G can use typically has shorter effective ranges than lower frequencies. Therefore, the 5G networks deployed on these higher frequencies may need more cell sites to deliver maximum performance.

Establishing new cell sites is a slow and expensive process. A widely discussed way to overcome this problem is to use public property—such as streetlights, traffic signals, utility poles, and bus shelters—as cell sites (WIA 2016). To date, however, the need to apply for construction and environmental permits at various levels of government, which have their own rules and fee structures, has created costs and delays (Forge et al. 2019).

As the owners of public infrastructure, governments can make that infrastructure available to 5G network builders at low cost and with an accelerated approval process. However, this appealing solution is not free of complications. For example, multiple layers of government and state-owned enterprises may be involved, public safety must be considered, and significant supporting infrastructure requirements (such as power supplies) may exist. Even once these issues are addressed, further implications may arise for market structure, competition, and shared access, among other matters.

Despite this complexity, governments are in a good position to improve operators' ability to establish sites quickly and cost-effectively. Reaching that goal must begin with policy and a clear strategic direction; in many cases, this work may require legislation and regulatory initiatives (refer to box 4.1).

BOX 4.1

Facilitating 5G deployment through infrastructure sharing in India

In 2018, India's Department of Telecommunications identified early deployment of efficient and pervasive 5G networks as a public good and sought ways to maximize the value offered by the new technology. A 5G Steering Committee report (August 2018) concluded that creating national 5G infrastructure would require massive additions of above-ground and below-ground infrastructure, ranging from backhaul radios, antennas, and towers to street furniture and ducts—an investment estimated at US$100 billion over 5–7 years.[a]

Because 5G densification implied 800–1,000 base stations per square kilometer, a facilitating policy for approvals and clearances was critical. Among the initiatives needed, the department recommended the creation of nationally uniform, strict guidelines for state and local governments to follow when issuing clearances. Streamlined online applications and prompt responses were identified as key improvements.[b] The department also recommended that the national government promulgate guidelines on smart infrastructure for state and local governments to use, with the goal being to advance uniform, practical adoption of infrastructure sharing.[c]

a. India's 5G Steering Committee Report (v26) is available on the Department of Telecommunications website: https://dot.gov.in/sites/default/files/5G%20Steering%20Committee%20report%20v%2026.pdf.

b. India's 5G Steering Committee Report (v26) is available on the Department of Telecommunications website: https://dot.gov.in/sites/default/files/5G%20Steering%20Committee%20report%20v%2026.pdf.

c. The Federal Communications Commission (FCC) in the United States implemented a similar approach to standardize procedures and fees for infrastructure deployment at the subnational level. It survived a lawsuit brought forward by dozens of city governments in which the US Court of Appeals ultimately supported the FCC's establishment of limits to fees charged to operators and required subnational governments to justify the levels of fees charged. Refer to the *City of Portland v. United States*, 2018, court opinion: https://cdn.ca9.uscourts.gov/datastore/opinions/2020/08/12/18-72689.pdf.

Backhaul

Backhaul denotes the connections that carry data from cell sites back to the rest of a network. As discussed in chapter 1, the technical specifications of backhaul links have a major impact on the overall performance of 5G. In particular, the type of backhaul available to connect sites will affect operators' decisions about implementing standalone versus non-standalone versions of 5G and, therefore, the capabilities of the 5G networks for end users.

Most mobile network sites currently use wireless technologies for backhaul (GSM Association 2019b). For 5G to deliver its full potential, these sites must be upgraded to fiber networks, which can provide much higher bandwidth at lower cost. However, fiber networks are expensive to install and, therefore, represent a major component of the capital costs of a full standalone 5G network. This cost has a significant impact on the financial viability of 5G network deployment—and thus on the likely evolution of geographical coverage.

Regulation plays an important role in how backhaul is deployed and used. The structure of the market for fiber-optic network providers is a critical factor in the availability of fiber connections to operators. A liberalized licensing regime and open markets promote entry and competition, stimulating investment in infrastructure and reducing prices. Other types of regulatory measures can support this overall policy. For example, rights of way and access to roads and other public infrastructure are important factors in the cost of deployment of fiber networks. Infrastructure sharing and access regulations also affect the underlying business case for new fiber network deployment. Furthermore, "dig once" policies and mandatory infrastructure sharing must be implemented in a way that does not shut out all but the most dominant players.

Infrastructure and Spectrum Sharing

Infrastructure sharing is an essential factor in determining the commercial viability of a 5G network. Sharing base stations across multiple MNOs can significantly reduce costs and accelerate deployment. Incentives to share network infrastructure vary. In markets with an unbalanced structure—such as those with one dominant player—incentives for the dominant player to share infrastructure are weak, as sharing would mean giving up some of its competitive advantage. In cases in which the market is more evenly balanced, operators may be keen to share infrastructure to reduce costs. The role of independent tower (mast) companies is also significant in many developing countries, as these provide an indirect way for operators to share infrastructure (GSM Association 2012).

In addition, the role of the regulator in infrastructure sharing varies between countries. In cases in which operators are reluctant to share, the regulator's task is to encourage them to do so in a fair and reasonable way. In other situations, regulators may be more focused on ensuring that sharing arrangements do not facilitate anticompetitive outcomes.

5G is expected to have further implications for infrastructure sharing. The high capital costs of infrastructure will give operators a strong incentive to share to reduce costs, particularly in the early days of 5G deployment, where uncertainty remains about the 5G business case. Furthermore, the possible use of public infrastructure for 5G cell sites has implications for sharing, as arrangements must be made to ensure equal treatment of operators. Finally, 5G potentially provides opportunities for spectrum, radio access networks, and other types of sharing models that might be supported by the industry. Such innovations may provide benefits through lower deployment costs, but they may have implications for competition that regulators must consider.

Network Slicing and Integration of 5G Networks into Vertical Applications

Many countries have enacted legislation on net neutrality to prevent undue discrimination in how network operators handle traffic from different sources. 5G's accommodation of network slicing differentiates it from earlier generations and is potentially a positive attribute for developing countries because it opens possibilities for providing broadband in commercial and other contexts. However, network slicing in 5G may be inconsistent with some countries' legal and regulatory frameworks for net neutrality. In such cases, operators will find it difficult to exploit the full capabilities of 5G without changes in the regulatory regime.

A related issue is the integration of 5G networks into vertical applications in business and industry, which has the potential to affect developing countries significantly, as explored in chapter 2. 5G's suitability for firm-specific applications is a new feature in both technical and commercial terms. Although it has generated considerable interest from industry and the mobile industry alike, much uncertainty remains about how it will be implemented and how successful it will be as a business model. Existing regulatory frameworks were designed with the current market structure and business models in mind. Therefore, they may not have the required degree of flexibility to accommodate the use of 5G in vertical applications, particularly if innovation or experimentation is required.

Recommendations for Regulatory Reform

Recommendations for regulatory reform include regulating the establishment of 5G sites, regulating the fiber-optic backhaul, setting up an enabling regulatory environment for infrastructure sharing, and setting up an enabling regulatory environment to encourage innovation in 5G.

Regulate the Establishment of 5G Sites
Regulatory authorities could support the development of sites for 5G networks in various ways. For example, they could streamline local site approval and construction permit procedures. This problem affects all types of mobile

network sites, not just 5G, and it has proved to be a significant challenge for operators and regulators. If regulators could find a way to reduce the administrative burden operators face in establishing new sites or upgrading old ones, all mobile networks and, ultimately, customers would benefit.

The deployment of 5G presents specific challenges in relation to sites. Existing sites will require upgrades to accommodate 5G equipment, including additional infrastructure for technologies such as massive multiple input, multiple output wireless technology. Again, it is likely that new sites will be required if 5G is to meet its full potential.

One approach that regulators should consider is applying a more streamlined approval process to small cell sites, starting with a consistent definition of "small sites," ideally harmonized with the definitions used in other countries. Any sites meeting this definition would be subject to streamlined processes and procedures for obtaining permits. Guidelines or a code of good practice could also be used to speed up parts of the process, including the role of local regulatory authorities.

REGULATE THE FIBER-OPTIC BACKHAUL

A streamlined approach should be applied to building fiber networks for backhaul from cell sites. This work would reduce the cost of building the fiber-optic backhaul to 5G sites, thereby improving the technical performance of the network, accelerating deployment, and lowering costs.

Telecommunications regulatory authorities should explore working with energy and transport regulators to encourage cross-infrastructure sharing of excess fiber capacity for backhaul purposes. This work has the benefit of leveraging the significant investments made by utilities in deploying fiber for their own internal needs, much of which is often in idle capacity, which could be used for backhauling. For this approach to become a reality, it will be critical to raise awareness among other regulatory authorities of the value of commercializing excess fiber capacity.

Earlier generations of mobile technology in developing countries depended primarily on wireless backhaul. Achieving the maximum potential of 5G networks will require an increasing reliance on fiber-optic cable backhaul. However, in some countries, the state still owns most of the fiber networks, and the markets associated with fiber-optic networks may not be competitive. Regulators should focus on reducing regulatory barriers to entry and fostering competition in the fiber-optic network market, including backhaul. This work will boost investment and innovation while reducing prices.

SET UP AN ENABLING REGULATORY ENVIRONMENT FOR INFRASTRUCTURE SHARING

Regulators should provide a clear regulatory framework for infrastructure sharing to cut the cost of network deployment while avoiding arrangements that might affect competition. Rules and guidelines on infrastructure sharing should be developed to suit each country's geographical, economic, and market conditions and include a broad range of factors, such as the following:

- Regulatory rules on the treatment of requests for infrastructure sharing;

- Technical standards to support commercial negotiations on infrastructure sharing;

- Guidelines on how the regulator will assess infrastructure-sharing requests (such as specifying which types of sharing can be implemented without regulatory approval and which warrant closer scrutiny) and whether and how competition authorities will be involved;

- Transparency on the location and characteristics of infrastructure that can be shared;

- A differentiated approach to facilitate sharing by dominant operators (including those that are state owned) at appropriate prices and in a nondiscriminatory way; and

- A differentiated approach for rural versus urban areas and treatment of sharing under specific circumstances (for example, sites within buildings in commercially valuable areas).

Set up an Enabling Regulatory Environment to Encourage Innovation in 5G

Existing legislation and regulations on net neutrality may affect innovation in technical and business models based on the use of 5G network slicing. If so, those frameworks may limit the benefits of 5G in industry, transport, and other key areas of the economy.

This complex issue is likely to affect many stakeholders within the mobile ecosystem. At a minimum, regulators should be aware of the implications of current and pending legal and regulatory rules for innovation around 5G. They should also monitor related developments in other countries to gauge the optimal interaction between legal and regulatory frameworks and 5G development.

Regulators should consider introducing flexibility into regulatory rules to encourage innovation by mobile operators and vertically structured industries. This flexibility can be introduced in the form of a regulatory "sandbox" environment, which has been used in the financial sector and embraced by many countries to facilitate testing new technology.[9] These sandboxes support innovation but within predetermined limits. If such innovation is successful and has the potential to expand outside the sandbox, regulators can then consider amending the general regulations so that the benefits are felt on a wider basis. Traditionally, agile regulatory approaches have not been applied in the telecommunications sector, but they warrant consideration in the context of 5G because of the potential to stimulate technical and business innovation. Various countries are taking this approach,[10] and some are implementing sandboxes for 5G specifically, such as Korea[11] and Thailand.[12]

To test the viability of technical solutions and business models in a specific sector or geographical location, 5G testbeds are an option. They have been implemented in Brazil[13] and the United Kingdom[14] and were planned in India in October 2021.[15] 5G testbeds help engineers to understand performance in real-world applications in controlled environments through pilot studies. Performance metrics can include feasibility, cost, and any issues that may impact scaling.[16]

In the United Kingdom, the government announced support for 5G testbeds across several scenarios to stimulate market development, including spectrum sharing, tourism, manufacturing and robotics, air and rail transportation, health care, and connectivity in rural communities, including farming. Although testbeds for testing 5G technology and business models can be useful in some circumstances, they require the right conditions and resources, such as well-functioning institutions. The necessary combination of critical foundational elements is less likely to be found in developing countries, where addressing broader institutional governance issues should be prioritized—a subject discussed at length later in this chapter.

Once the institutional capacity is in place, the requirements and recommendations for setting up a 5G testbed as outlined by the Institute of Electrical and Electronics Engineers Future Networks platform include obtaining permission to set up a 5G testbed to use available radio frequency and then setting up a team of relevant researchers and expertise.[17] This issue underscores the need for collaborations among governments, industry, and, in particular, academic and research institutions to yield successful 5G testbed development and deployment.

The use of 5G in vertical industries may have cross-border regulatory implications. For example, high-capacity, low-latency applications of the internet of things using 5G may require international harmonization of standards and involve international roaming on a permanent or semi-permanent basis (such as for 5G network connected cars). These issues have arisen in earlier generations of mobile networks, but they may take on added importance given 5G's potential for deeper integration into vertically structured industries. Regulators in developing countries should carefully monitor international developments in this area and consider whether any amendments to the national regulatory framework would be required to support innovative applications.

INSTITUTIONAL GOVERNANCE AND CAPACITY BUILDING

The traditional model of regulation in the telecommunications sector has centered on licensed telecommunications network operators and service providers. Over time, the growth of the broader digital service sector and its gradual melding with adjacent sectors, such as media, have prompted institutional changes in countries like Singapore and the United Kingdom,[18] but the traditional model of regulation remains prevalent in many developing countries.

Redesign of Regulatory Authorities

The innovations that 5G could bring may result in a further reconsideration of the institutional design of regulatory authorities. 5G's integration into vertically structured industries, such as manufacturing, extractives, and transport, will potentially bring these sectors within the purview of the telecommunications regulator. Conversely, telecommunications operators that were traditionally regulated by a single regulatory authority may be required to interact with several sector authorities and face increasing constraints from horizontal legislation (including data protection and cybersecurity).

For example, digital services delivered by 5G networks—such as autonomous driving, remote health monitoring, and autonomous drones—intersect with regulatory systems beyond those governing telecommunications. Such services would involve regulators of utilities, traffic safety, financial services, health care, and air safety on the local, national, and international levels.

Cooperation and Coordination among Regulatory Authorities and Government Institutions

For 5G to achieve its full potential through applications in other sectors, a new approach to regulation may be needed. At a minimum, closer cooperation and coordination between regulatory authorities and other government institutions will be needed to ensure a coherent approach to 5G development. Some countries may consider redesigning regulatory institutions so that the necessary coordination is embedded within the government.[19]

Recommendation for Institutional Governance and Capacity Building: Establish a Cross-Government Strategy

Governments of developing countries should consider developing a 5G strategy that includes a cross-government approach to policy and regulation. The strategy should consider the potential of 5G networks to interact with other sectors and areas of government and recognize the need for a coherent approach to regulation.

The optimal approach to institutional design is likely to vary by country. No single approach will work everywhere. Therefore, governments should consider a country's starting conditions and overall economic strategy when developing 5G strategy and plans for institutional reforms.

The importance of using this fine-tuned approach should not be underestimated. Much of the potential benefit of 5G comes from its ability to integrate wireless broadband into production processes, service delivery networks, and other types of public and private activities. A failure to ensure a coordinated approach across multiple government institutions would limit the potential of 5G to deliver desirable outcomes in developing countries.

CONCLUSION

This chapter presented some of the most pressing challenges that policy makers will face in developing a 5G ecosystem. These challenges include limited spectrum availability, regulatory frameworks ill-suited for 5G deployment, insufficient institutional capacity, weak competition policies, and overly restrictive licensing regimes. These issues are not exclusive to developing countries, as they affect most countries on the path to 5G. An important lesson for policy makers is that institutional capacity, forward thinking, and public-private partnerships will improve any country's readiness to deploy the latest generation of mobile technologies.

The recommendations presented in this chapter highlight the need to devise national strategies, reduce barriers for stakeholders in the 5G ecosystem, and be ready to adapt to a rapidly evolving landscape. Implemented now, the right strategy can lay the foundation for the adoption and diffusion of 5G technologies and allow it to produce maximum welfare gains. Appendix B provides a selection of extant national 5G strategy documents that can be referenced to understand how other countries have approached policy planning, regulatory matters, and cross-government coordination. In the short term, countries should consider initiating public consultations with key stakeholders in the 5G ecosystem—including providers, industry, and users alike—to develop sensitive, informed, and coordinated policy responses. Insights from these consultations should inform long-term national strategies for 5G, broadband, spectrum, and cybersecurity—all leading to a 5G-enabled economy.

As a reference for decision-makers, box 4.2 provides a roadmap for the path to 5G, informed by the key messages and topics highlighted throughout this book.

BOX 4.2

A roadmap for the path to 5G in developing countries

Every country is different and, as such, will need its own roadmap toward 5G. 5G can offer many potential benefits but has significant costs associated with it, and the value of its benefits will depend on a country's specifics. This roadmap highlights three main issues that governments and policy makers should consider in preparation for 5G in their countries:

1. Assessing the need for 5G and its potential benefits;

2. Developing a strategy to deliver optimal wireless connectivity; and

3. Improving infrastructure and regulations, which will bring benefits regardless of the approach adopted.

(continued)

BOX 4.2 *(continued)*

Assessing Need

- The balance of key needs in terms of coverage, capacity, and performance features;
- The degree to which wireless is a key element for delivering broadband;
- The verticals or sectors for which 5G features are likely to become important; and
- Any national strategies or plans that might impact a 5G strategy.

This assessment can lead to a clear understanding of where 5G can add value and, conversely, where other approaches, such as increased fiber deployment or enhanced 4G coverage, are more appropriate.

Developing a Strategy

- Where 5G should be deployed based on a needs assessment,
- The availability of spectrum for 5G,
- The best balance between competition and optimization of network costs,
- The extent to which intervention is required to deliver this strategy, and
- How best to convene industry players and other stakeholders to collaborate on supporting the launch of financially viable 5G services.

This strategy should lead to a plan to enable the delivery of 5G in areas where it adds value.

Improving Infrastructure and Regulations

- More extensive fiber deployments;
- A better-equipped regulator able to deploy leading-edge spectrum management tools;
- Establishment of regulatory flexibility to support an enabling environment for the sector;
- Preparation of spectrum, including removal of incumbents and spectrum for private networks, as well as any generation-specific restrictions on existing licenses;
- Removal of barriers to mobile deployment, including bureaucratic and slow site access and onerous regulatory requirements; and
- Support for a national innovation ecosystem to spur innovation in new use cases and services enabled by 5G and future advanced mobile networks, understanding that new use cases are difficult to predict.

NOTES

1. Calculations based on data from the GSM Association Spectrum Navigator database, accessed September 3, 2021.
2. A tiered framework is already used for citizens' broadband in the 3.5 GHz band (3,550–3,700 MHz), which is suitable for deployment of 4G and

5G. Refer to https://www.fcc.gov/wireless/bureau-divisions/mobility
-division/35-ghz-band/35-ghz-band-overview.

3. Examples are from Cullen International (2021).

4. GSM Association recommends 3–5 years of planning for a spectrum roadmap;
for more details, refer to https://www.gsma.com/spectrum/wp-content
/uploads/2017/11/4-Day-1-Session-1-How-to-design-a-spectrum-roadmap
-Richard-Marsden.pdf and https://www.gsma.com/spectrum/wp-content
/uploads/2019/03/Laurent-Bodusseau-Spectrum-Roadmap.pdf.

5. The United Kingdom's 2021 Spectrum Strategy Statement is available at https://
www.ofcom.org.uk/__data/assets/pdf_file/0017/222173/spectrum-strategy
-statement.pdf.

6. For more details on "use-or-share" clauses, refer to Calabrese (2021).

7. For more information on dynamic spectrum-sharing regulations, refer to the
Dynamic Spectrum Alliance website: http://dynamicspectrumalliance.org
/regulations/.

8. For an example of a framework designed to facilitate sharing in local areas in
the United Kingdom, refer to the Ofcom website: https://www.ofcom.org.uk
/manage-your-licence/radiocommunication-licences/shared-access.

9. For a collection of regulatory sandboxes around the world, refer to https://
dfsobservatory.com/regulatory-sandbox.

10. For a review of how emerging technologies are driving this new regulatory
approach across various sectors, including communications, refer to OECD
(2020).

11. For details on Korea's regulatory sandbox for 5G, refer to the notice on the
Ministry of Science and ICT website: https://english.msit.go.kr/eng/bbs/view
.do?sCode=eng&mId=7&mPid=2&pageIndex=&bbsSeqNo=44&nttSeqNo=137
&searchOpt=&searchTxt=.

12. For details on Thailand's regulatory sandbox in the telecommunications sector,
refer to Malisuwan, Chayawan, and Kaewphanuekrungsi (2020).

13. For details on Brazil's 5G testbed initiatives, refer to https://ieeexplore.ieee.org
/document/8696721/.

14. For details on the United Kingdom's 5G Testbeds and Trials Programme, refer to
https://www.gov.uk/guidance/5g-testbeds-and-trials-programme.

15. For details on India's planned 5G testbed, refer to https://economictimes
.indiatimes.com/industry/telecom/telecom-news/expected-by-oct-5g-test-bed-to
-boost-telecom-technology/articleshow/81245539.cms.

16. For an explanation of 5G testbeds, refer to https://futurenetworks.ieee.org
/topics/5g-testbed.

17. For tools and resources on the development of 5G testbeds, refer to https://future
networks.ieee.org/topics/5g-testbed.

18. For example, the creation of Ofcom in the United Kingdom in 2003 was
designed to replace five regulatory bodies that had traditionally overseen media
and communications channels: the Broadcasting Standards Commission, the
Independent Television Commission, the Office of Telecommunications, the
Radio Authority, and the Radiocommunications Agency. For a review of Ofcom's
remit, refer to https://www.ofcom.org.uk/about-ofcom/what-is-ofcom.

19. For example, some countries already have network sector regulation (including
telecom) and antitrust under one institution, such as Australia, the Netherlands,
and Spain, which could facilitate coordination.

BIBLIOGRAPHY

Government Resources

[Brazil] Federal Constituição. 2019. "Diário Oficial da União. Law Nº 13.874." http://www .in.gov.br/en/web/dou/-/lei-n-13.874-de-20-de-setembro-de-2019-217365826.

[Brazil] Presidência da República Secretaria-Geral Subchefia para Assuntos Jurídicos. 2015. "Law Nº 13.116." http://www.planalto.gov.br/ccivil_03/_Ato2015-2018/2015 /Lei/L13116.htm.

[China] ICT Academy. 2016. *Broadband China Strategy and Its Implementation*. Ministry of Industry and Information Technology. https://www.unescap.org/sites/default /files/Broadband%20China%20Strategy.pdf.

[China] Ministry of Industry and Information Technology (MIIT). 2019. "工业和 信息化部印发《5G+工业互联网"512工程推进方案》 [The Ministry of Industry and Information Technology Issued the '5G+Industrial Internet' 512 Project Promotion Plan]." Press Release, November 24. http://www.cac.gov.cn/2019-11 /24/c_1576133540276534.htm.

[China] Ministry of Industry and Information Technology (MIIT). 2020. "Notice Concerning Promoting the Accelerated Development of 5G." MIIT Communication No. (2020) 49. English translation available from the China Copyright and Media Blog edited by Rogier Creemers, March 26. https:// chinacopyrightandmedia.wordpress.com/2020/03/24/notice-concerning -promoting-the-accelerated-development-of-5g/.

[China] Shenzhen Municipal People's Government. 2019. "深圳市人民政府印发 关于率先实现5G基础设施全覆盖及促进5G产业高质量发展若干措施的通知 [The Shenzhen Municipal People's Government Issued a Notice on Several Measures to Take the Lead in Achieving Full Coverage of 5G Infrastructure and Promoting the High-Quality Development of the 5G Industry]." Press Release, September 11. http://www.sz.gov.cn/zfgb/2019/gb1115/content/post_4951288.html.

[Colombia] El Congreso de Colombia. 2019a. "Law 1955 of May 2019: El Plan Nacional de Desarrollo 2018–2022." May 25. https://colaboracion.dnp.gov.co/CDT/Prensa /Ley1955-PlanNacionaldeDesarrollo-pacto-por-colombia-pacto-por-la-equidad.pdf.

[Colombia] El Congreso de Colombia. 2019b. "Ley 1978 de 2019: Por la cual se moderniza el Sector de las Tecnologías de la Información y las Comunicaciones -TIC, se distribuyen competencias, se crea un Regulador Único y se dictan otras disposiciones." June 25. https://www.funcionpublica.gov.co/eva/gestornormativo/norma.php?i=98210.

[Colombia] Ministerio de Tecnologías de la Información y las Comunicaciones de Colombia. 2020a. "MinTIC publica resoluciones para uso del espectro destinado a pruebas técnicas." March 9. https://www.mintic.gov.co/portal /inicio/Sala-de-Prensa/Noticias/126094:MinTIC-publica-resoluciones -para-uso-del-espectro-destinado-a-pruebas-tecnicas.

[Colombia] Ministerio de Tecnologías de la Información y las Comunicaciones de Colombia. 2020b. "Resolución Número 000467 del 9 de marzo de 2020." March 9. https://www.mintic.gov.co/portal/604/articles-126094_recurso_2.pdf.

[Colombia] Ministerio de Tecnologías de la Información y las Comunicaciones de Colombia. 2020c. "Resolución número 000468 del 9 de marzo de 2020." March 9. https://www.mintic.gov.co/portal/604/articles-126094_recurso_1.pdf.

[Colombia] Ministerio de Tecnologías de la Información y las Comunicaciones de Colombia. 2020d. "Cancelado el Facebook Live para el lanzamiento de la convocatoria de pilotos 5G." March 16. https://www.mintic.gov.co/portal /inicio/Sala-de-Prensa/Noticias/126185:Cancelado-el-Facebook-Live-para-el -lanzamiento-de-la-convocatoria-de-pilotos-5G.

[Egypt] Ministry of Communications and Information Technology. 2014. "Unified License System Approval." Cairo, April 2. http://mcit.gov.eg/Media_Center /Press_Room/Press_Releases/3010.

[European Commission] Communications Committee (COCOM), Working Group on 5G. 2018. "Report on the Exchange of Best Practices Concerning National Broadband Strategies and 5G 'Path-to-Deployment'." Working Document COCOM18-06REV-2. https://ec.europa.eu/newsroom/dae/document.cfm?doc_id=55605.

[Germany] Bundesnetzagentur (BNetzA). 2018. "President's Chamber Decision of 14 May 2018 on the 77Order for and Choice of Proceedings for the Award of Spectrum in the 2 GHz and 3.6 GHz Bands for Mobile/Fixed Communication Networks (MFCN)." https://www.bundesnetzagentur.de/SharedDocs/Downloads/EN/Areas/Telecommunications/Companies/TelecomRegulation/FrequencyManagement/ElectronicCommunicationsServices/FrequencyAward2018/20180613_Decision_I_II.pdf?__blob=publicationFile&v=2.

[Germany] Bundesnetzagentur (BNetzA). 2020. "Numerous Frequency Assignments for 5G Campus Networks." https://www.bundesnetzagentur.de/SharedDocs/Downloads/EN/BNetzA/PressSection/PressReleases/2020/202009021_5GCampusNetworks.pdf?__blob=publicationFile&v=2.

[India] Department of Telecommunications (DoT). 2016. "Indian Telegraph Rights of Way Rules." *Gazette of India.* https://dot.gov.in/sites/default/files/2016_11_18%20RoW%20Policy.pdf.

[Singapore] Infocomm Media Development Authority (IMDA). 2019. "Policy for the Fifth-Generation (5G) Mobile Networks and Services in Singapore." https://www.imda.gov.sg/-/media/Imda/Files/Regulation-Licensing-and-Consultations/Consultations/Consultation-Papers/Second-.

[South Africa] Independent Communications Authority of South Africa (ICASA). 2020. "Temporary Radio Spectrum Issued to Qualifying Applicants in an Effort to Deal with COVID-19 Communication Challenges." April 17. https://www.icasa.org.za/news/2020/temporary-radio-frequency-spectrum-issued-to-qualifying-applicants-in-an-effort-to-deal-with-covid-19-communication-challenges.

[US] Broadband Deployment Advisory Committee (BDAC). 2018. "Model Code for Municipalities." Federal Communications Commission. https://www.fcc.gov/sites/default/files/bdac-07-2627-2018-model-code-for-municipalities-approved-rec.pdf.

[US] Code of Federal Regulations (CFR). 2020. "Section 1.1312: Facilities with No Preconstruction Authorization Requirements." https://www.govregs.com/regulations/expand/title47_chapterI_part1_subpartI_section1.1312.

[US] Federal Communications Commission (FCC). 2018a. "Strategic Plan 2018–2022." https://docs.fcc.gov/public/attachments/DOC-349143A1.pdf.

[US] Federal Communications Commission (FCC). 2018b. "The FCC's 5G FAST Plan." https://www.fcc.gov/5G.

[US] Federal Communications Commission (FCC). 2019. "Notice of Proposed Rulemaking: In the Matter of Updating the Commission's Rule for Over-the-Air Reception Devices." WT Docket No. 19-71, Adopted April 12. https://docs.fcc.gov/public/attachments/FCC-19-.

[US] Federal Communications Commission (FCC). 2020. "FCC Proposes the 5G Fund for Rural America." Washington, DC, April 23. https://docs.fcc.gov/public/attachments/DOC-363946A1.pdf.

[US] National Telecommunications and Information Administration (NTIA). 2020. "Request for Comments on the National Strategy to Secure 5G Implementation Plan." Department of Commerce, May 28. https://www.ntia.doc.gov/federal-register-notice/2020/request-comments-national-strategy-secure-5g-implementation-plan.

[US] White House. 2019. "Remarks by President Trump on United States 5G Deployment." Press Release. https://www.whitehouse.gov/briefings-statements/remarks-president-trump-united-states-5g-deployment/.

[US] White House. 2020. "National Strategy to Secure 5G." https://www.whitehouse.gov/wp-content/uploads/2020/03/National-Strategy-5G-Final.pdf.

[Viet Nam] Ministry of Planning and Investment. 2019. "Resolution No. 52-NQ/TW on Guidelines and Policies to Actively Participate in the Fourth Industrial Revolution." Hanoi, September 27. https://thuvienphapluat.vn/van-ban/Dau-tu/Nghi-quyet-52-NQ-TW-2019-chinh-sach-chu-dong-tham-gia-cuoc-Cach-mang-cong-nghiep-lan-thu-tu-425113.aspx.

General Resources

Amazon. 2013. *Broadband China Strategy and Implementation Plan.* https://www.amazon.com/Broadband-China-strategy-implementation-Chinese/dp/7010125031.

Belarusian Telegraph Agency. 2019. "Plans to Develop 5G Technology in Belarus Fleshed Out." Minsk, April 9. https://eng.belta.by/society/view/plans-to-develop-5g-technology-in-belarus-fleshed-out-120134-2019/.

Bell, Pete. 2018. *Italian 5G Auction Sees High Price Tags, Raised Eyebrows.* October 15. https://blog.telegeography.com/italian-5g-auction-causes-concern.

Benghozi, Pierre-Jean. 2018. "French Perspective on Spectrum Issues." Madrid, September 14. https://www.ucm.es/data/cont/media/www/pag-115737/Pierre-Jean%20Benghozi.%20Spectrum%20Madrid%202018%20PJB.pdf.

BEREC. 2018. *BEREC Report on Infrastructure Sharing.* https://berec.europa.eu/eng/document_register/subject_matter/berec/download/0/8164-berec-report-on-infrastructure-sharing_0.pdf.

Blackman, Colin, and Lara Srivastava. 2011. "Telecommunications Regulation Handbook." *World Bank and the International Telecommunications Union,* Washington, DC: World Bank. http://hdl.handle.net/10986/13278.

Bulgarian National Radio (BNR). 2020. "There Are No Functioning 5G Networks in Bulgaria at This Time: Transport Minister." May 20. https://bnr.bg/en/post/101278983/there-are-no-functioning-5g-networks-in-bulgaria-at-this-time-transport-minister.

Bushberg, J. T., C. K. Chou, K. R. Foster, R. Kavet, D. P. Maxson, R. A. Tell, and M. C. Ziskin. 2020. "COMAR Technical Information Statement: Health and Safety Issues Concerning Exposure of the General Public to Electromagnetic Energy from 5G Wireless Communications Networks." *Health Physics,* June 22. https://pdfs.journals.lww.com/health-physics/9000/00000/IEEE_Committee_on_Man_and_Radiation_Comar.99768.pdf.

Calabrese, Michael. 2021. "Use It or Share It: A New Default Policy for Spectrum Management." TPRC48: The 48th Research Conference on Communication: Information and Internet Policy. doi:10.2139/ssrn.3762098.

Clark, Robert. 2019. "Operators Starting to Face Up to 5G Power Cost." *Light Reading,* October 30. https://www.lightreading.com/asia-pacific/operators-starting-to-face-up-to-5g-power-cost-/d/d-id/755255#:~:text=The%20power%20consumption%20of%205G,a%204G%20basestation%2C%20it%20says.&text=Huawei%20estimates%20that%20by%202026,million%20small%20cells%20u.

Cullen International. 2021. "Who Is Deploying Private Mobile Networks, and Why?" Cullen International, Brussels, Belgium.

de Villiers, James. 2019. "Why Telkom Will Not Yet Invest in 5G—And When 5G Will Be Available at Cell C, Vodacom, MTN, and Rain." *Business Insider—South Africa,* May 6. https://www.businessinsider.co.za/south-africa-5g-rollout-when-telkom-cell-c-vodacom-mtn-rain-2019-4.

Deloitte for GSMA. 2015. "Digital Inclusion and Mobile Sector Taxation in Tanzania." https://www.gsma.com/publicpolicy/wp-content/uploads/2016/09/GSMA2015_Report_DigitalInclusionAndMobileSectorTaxationInTanzania.pdf.

Digital Policy Alliance. n.d. "About Digital Infrastructure Group." https://www .dpalliance.org.uk/about-digital-infrastructure-group/.

European 5G Observatory. 2019. "Japan Assigns 5G Spectrum to Four Operators." https://5gobservatory.eu/japan-assigns-5g-spectrum-to-four-operators/.

European Commission (EC). 2012. *Communication from the Commission ... Promoting the Shared Use of Radio Spectrum Resources in the Internal Market.* COM (2012) 478 Final, September 3. https://ec.europa.eu/digital-single-market/sites/digital-agenda/files /com-.

European Commission (EC). 2016a. "5G for Europe: An Action Plan. COM (2016) 588 Final." https://ec.europa.eu/transparency/regdoc/rep/1/2016/EN/1-2016 -588-EN-F1-1.PDF.

European Commission (EC). 2016b. *Communication from the Commission ... Connectivity for a Competitive Digital Single Market—Towards a European Gigabit Society.* COM (2016) 0587 Final, September 14. https://eur-lex.europa.eu/legal-content/en /TXT/?uri=CELEX%3A52016DC0587.

European Commission (EC). 2020. "Commission Implementing Regulation of 30 June 2020 on Specifying the Characteristics of Small-Area Wireless Access Points Pursuant to Article 57 Paragraph 2 of Directive (EU) 2018/1972." https://ec.europa .eu/newsroom/dae.

European Communications Committee (ECC). 2014. *ECC Report 205: Licensed Shared Access.* February. https://www.ecodocdb.dk/download/baa4087d-e404 /ECCREP205.PDF.

Forge, Simon, Robert Horvitz, and Colin Blackman. 2014. *Is Commercial Cellular Suitable for Mission Critical Broadband?* European Commission. https://op.europa.eu /en/publication-detail/-/publication/246bc6ec-6251-40cb-aab6-748ae316e56d.

Forge, Simon, Robert Horvitz, Colin Blackman, and Erik Bohlin. 2019. *Light Deployment Regime for Small-Area Wireless Access Points (SAWAPs).* European Commission. https://op.europa.eu/en/publication-detail/-/publication/463e2d3d-1d8f -11ea-95ab-01aa75ed71a1/language-en.

Gallagher, Jill C., and Michael E. DeVine. 2019. "Fifth-Generation (5G) Telecommunications Technologies: Issues for Congress." Congressional Research Service, CRS Report R45485. http://www.hsdl.org/?view&did=821493.

García Zaballos, Antonio, Enrique Iglesias Rodríguez, Kyoung Woo Kim, and Soontae Park. 2020. *5G: The Driver for the Next-Generation Digital Society in Latin America and the Caribbean.* Washington, DC: Inter-American Development Bank. https:// publications.iadb.org/publications/english/document/5G_The_Driver_for_the _Next-Generation_Digital_Society_in_Latin_America_and_the_Caribbean.pdf.

Global Mobile Suppliers Association (GSA). 2020. "Private LTE & 5G Networks Report February." https://gsacom.com/paper/private-lte-5g-networks-report -february-2020/.

Government of New York City (NYC). 2020. "Mobile Telecommunications Franchises FAQs." https://www1.nyc.gov/site/doitt/business/mobile-telecom-faqs.page.

Government of the Netherlands. 2016. "Netherlands Radio Spectrum Policy Memorandum." Ministry of Economic Affairs and Climate Policy. https:// www.government.nl/documents/reports/2017/03/07/radio-spectrum -policy-memorandum-2016.

Government of the Republic of Korea. 2017. *People-Centered Plan for the Fourth Industrial Revolution to Promote Innovative Growth (I-KOREA 4.0).* https://www.4th-ir .go.kr/article/detail/220.

Government of the Republic of Korea. 2019. *5G+ Strategy: To Realize Innovative Growth.* https://www.msit.go.kr/cms/english/pl/policies2/__icsFiles/afieldfile/2020/01 /20/5G%20plus%20Strategy%20to%20Realize%20Innovative%20Growth.pdf.

Greensill. 2019. *Financing the Future of 5G.* October 21. https://www.greensill.com /whitepapers/financing-the-future-of-5g/.

GSM Association. 2012. *Mobile Infrastructure Sharing.* September. https://www.gsma .com/publicpolicy/wp-content/uploads/2012/09/Mobile-Infrastructure-sharing .pdf.

GSM Association. 2017. *Spectrum Pricing: GSMA Public Policy Position.* September. https://www.gsma.com/spectrum/wp-content/uploads/2018/12/spectrum _pricing_positioning_2017.pdf.

GSM Association. 2018a. *Study on Socio-Economic Benefits of 5G Services Provided in mmWave Bands.* December. https://www.gsma.com/spectrum/wp-content/uploads/2019/10 /mmWave-5G-benefits.pdf.

GSM Association. 2018b. *Spectrum Pricing in Developing Countries: Evidence to Support Better and Affordable Mobile Services.* https://data.gsmaintelligence.com/api-web /v2/research-file-download?id=33292319&file=Spectrum%20pricing%20in%20 developing%20countries.pdf.

GSM Association. 2019a. *The 5G Guide: A Reference for Operators.* April. https://www .gsma.com/wp-content/uploads/2019/04/The-5G-Guide_GSMA_2019_04_29 _compressed.pdf.

GSM Association. 2019b. *Mobile Backhaul: An Overview.* June 19. https://www.gsma .com/futurenetworks/wiki/mobile-backhaul-an-overview/.

GSM Association. 2020. *The Mobile Economy 2020.* https://www.gsma.com /mobileeconomy/wp-content/uploads/2020/03/GSMA_MobileEconomy2020 _Global.pdf.

Harb, Robbie. 2020. "Korea Announces Tech New Deal to Stoke Post-Coronavirus Economy." *The Register.* https://www.theregister.co.uk/2020/05/11/south_korea _tech_new_deal/.

Hardesty, Linda. 2019. "China's Carriers to Build a Shared 5G Network." *FierceWireless.* https://www.fiercewireless.com/5g/china-s-carriers-to-build-a-shared -5g-network.

Hazlett, Thomas W., and Roberto E. Muñoz. 2009. "Welfare Analysis of Spectrum Allocation Policies." *The RAND Journal of Economics* 40 (3): 424–54. doi:10.1111/j.1756-2171.2009.00072.x.

ICNIRP. n.d. "5G Radiofrequency—RF EMF." https://www.icnirp.org/en/applications /5g/index.html.

International Telecommunication Union (ITU-R). 2001. *Recommendation SM.1132-2 General Principles and Methods for Sharing between Radiocommunication Services or between Radio Stations.* July. https://www.itu.int/rec/R-REC-SM.1132-2-200107-I/en.

International Telecommunication Union (ITU-R). 2003. *Recommendation SM.1603 Spectrum Redeployment as a Method of National Spectrum Management.* https://www.itu .int/rec/R-REC-SM.1603-2-201408-I/en.

International Telecommunication Union (ITU-T). 2018a. *Recommendations: The Impact of RF-EMF Exposure Limits Stricter than the ICNIRP or IEEE Guidelines on 4G and 5G Mobile Network Deployment.* Supplement 14 to ITU-T K-series. May 5. https://www.itu.int /rec/dologin_pub.asp?lang=e&id=T-REC-K.Sup14-201909-I!!PDF-E&type=items.

International Telecommunication Union (ITU-R). 2018b. *Recommendation ITU-T K.52 Guidance on Complying with Limits for Human Exposure to Electromagnetic Fields.* January. https://www.itu.int/rec/dologin_pub.asp?lang=e&id=T-REC-K.52-201801 -I!!PDF-E&type=items#:~:text=Summary-,Recommendation%20ITU%2DT%20 K.,to%20electromagnetic%20fields%20(EMFs).

JQK News. 2020. "Policy Support 5G Construction Is Expected to Drive Investment of More than 3.5 Trillion Yuan in the Next Five Years." June 8. https://www .jqknews.com/news/481197-Policy_support_5g_construction_is_expected_to _drive_investment_of_more_than_35_trillion_yuan_in_the_next_five_years.html.

Khan, Danish, and Tina Gurnaney. 2017. "Right of Way Rules: The Effects of Implementation Delay on India's Telecom Industry." *Economic Times of India.* https:// telecom.economictimes.indiatimes.com/news/right-of-way-rules-the-effects-of -implementation-delay-on-india-telecom-industry/59855964.

Korean Communications Commission (KCC). 1997. "Radio Waves Act No. 5454, Dec. 13, 1997." http://elaw.klri.re.kr/eng_service/lawView.do?hseq=7162&lang=ENG.

Kuroda, Toshifumi, and Maria del Pilar Baquero Forero. 2017. "The Effects of Spectrum Allocation Mechanisms on Market Outcomes: Auctions versus Beauty Contests." *Telecommunications Policy* 41 (5–6): 341–54. doi:10.1016/j.telpol.2017.01.006.

Labovitz, Craig. 2020. "Early Effects of COVID-19 Lockdowns on Service Provider Networks: The Networks Soldier On!" *Nokia*, March 20. https://www.nokia.com/blog/early-effects-covid-19-lockdowns-service-provider-networks-networks-soldier/.

Lee, Paul, Mark Casey, and Craig Wigginton. 2019. "Private 5G Networks: Enterprise Untethered." *Deloitte Insights.* https://www2.deloitte.com/us/en/insights/industry/technology/technology-media-and-telecom-predictions/2020/private-5g-networks.html.

Malisuwan, S., C. Chayawan, and W. Kaewphanuekrungsi. 2020. "Establishment of Regulatory Sandbox: A Case Study of Thailand's Regulatory Sandbox." *International Journal of the Computer, the Internet and Management* 28 (1). http://www.ijcim.th.org/past_editions/2020V28N1/28n1Page84.pdf.

Matheson, Thornton, and Patrick Petit. 2017. "Taxing Telecommunications in Developing Countries." White Paper WP/17/247, International Monetary Fund. https://www.imf.org/en/Publications/WP/Issues/2017/11/15/Taxing-Telecommunications-in-Developing-Countries-45349.

McKinsey Global Institute. 2020. *Connected World: An Evolution in Connectivity Beyond the 5G Revolution.* Discussion Paper. February. https://www.mckinsey.com/~/media/mckinsey/industries/technology%20media%20and%20telecommunications/telecommunications/our%.

McLeod, Duncan. 2017. "SA's First 5G Network to Go Live Next Month." *Tech Central*, October 11. https://techcentral.co.za/sas-first-5g-network-go-live-next-month/77470/.

McLeod, Duncan. 2019. "Icasa Files Suit against Communications Minister Over Budget." *Tech Central*, April 29. https://techcentral.co.za/icasa-files-suit-against-communications-minister-over-budget/89265/.

Meddour, Djamal-Eddine, Tinku Rasheed, and Yvon Gourhant. 2011. "On the Role of Infrastructure Sharing for Mobile Network Operators in Emerging Markets." *Computer Networks* 55. https://www.sciencedirect.com/science/article/abs/pii/S1389128611000776.

MyBroadband. 2020. "MTN to Launch New 5G Network in June." *MyBroadband*, June 2. https://mybroadband.co.za/news/5g/354609-mtn-to-launch-new-5g-network-in-june.html.

Nigerian Communications Commission (NCC). 2020. "Press Statement: No License Has Been Issued for 5G in Nigeria." April 4. https://www.ncc.gov.ng/media-centre/news-headlines/812-press-statement-no-licence-has-been-issued-for-5g-in-nigeria.

Nikkei Asian Review. 2019. "Japan Weighs 15% Tax Credit to Accelerate 5G Investment." December 11. https://asia.nikkei.com/Spotlight/5G-networks/Japan-weighs-15-tax-credit-to-accelerate-5G-investment.

OECD (Organisation for Economic Co-Operation and Development). 2020. *Shaping the Future of Regulators: The Impact of Emerging Technologies on Economic Regulators.* Paris: OECD Publishing. doi:10.1787/db481aa3-en.

Radio Spectrum Policy Group (RSPG). 2013. "RSPG Opinion on Licensed Shared Access." RSPG13-538. https://circabc.europa.eu/sd/d/3958ecef-c25e-4e4f-8e3b-469d1db6bc07/RSPG13-538_RSPG-Opinion-on-LSA%20.pdf.

Ramaphosa, Cyril. 2019. "State of the Union Address [Speech]." *Mail & Guardian*, Cape Town, June 20. https://mg.co.za/article/2019-06-20-read-it-in-full-ramaphosas-state-of-the-nation-address/.

Sahi, Nokhaiz. 2020. "Ministry Holds 1st Meeting of Advisory Committee for 5G Planning in Pakistan." *The Nation.* https://nation.com.pk/10-Mar-2020/ministry -holds-1st-meeting-of-advisory-committee-for-5g-planning-in-pakistan.

Small Cell Forum (SCF). 2017. "Small Cell Forum Unveils Operator Research Showing Accelerating Densification and Enterprise Deployments on Road to 5G." https://www.smallcellforum.org/press-releases/small-cell-forum-unveils-operator -research-showing-accelerating-densification-enterprise-deployments-road-5g/.

Small Cell Forum (SCF). 2018. *Small Cells Market Status Report Document 050.10.03.* December. https://scf.io/en/documents/050_-_Small_cells_market_status_report _December_2018.php.

Song, Su-hyun. 2018. *Auction for 5G Frequency Spectrums Set on June 15.* May 3. http:// www.koreaherald.com/view.php?ud=20180503000713.

Stuckmann, Peter. 2018. "Proud to Announce Our New Checklist for National 5G Strategies." *Shaping Europe's Digital Future Blog.* https://ec.europa.eu/digital-single -market/en/blogposts/proud-announce-our-new-checklist-national-5g-strategies.

Tang, Frank. 2020. "Coronavirus: China, Xi Jinping put 5G Technology on Top of Huge Spending Plans to Salvage Economy." *South China Morning Post.* https:// www.scmp.com/economy/china-economy/article/3065229/coronavirus -china-xi-jinping-put-5g-technology-top-huge.

Tech Central. 2017. "Vodacom to Begin 5G Trials with Nokia." November 7. https:// techcentral.co.za/vodacom-begin-5g-trials-nokia/78040/.

Tech Central. 2020a. "MTN 5G Launch: Everything You Need to Know." June 30. https://techcentral.co.za/mtn-5g-launch-everything-you-need-to-know/99224/.

Tech Central. 2020b. "Interview with Giovanni Chiarelli on MTN South Africa's 5G Launch." June. https://www.youtube.com/watch?v=wz-n9uyH9-4.

Telebrasil. 2018. *Ranking Cidades Amigas.* 3rd ed. http://telebrasil.org.br/component /docman/doc_download/1888-ranking-cidades-amigas-da-internet-2018?Itemid=.

Telecomlead. 2020. *Finland Raises 21 mn Euros from Second 5G Spectrum Auction.* June 8. https://www.telecomlead.com/5g/finland-raises-21-mn-euros-from-second-5g -spectrum-auction-95474.

Tomás, Juan Pedro. 2019. "Ericsson Wins 5G Network Contract in South Africa." *RCR Wireless.* November 14. https://www.rcrwireless.com/20191114/5g/ericsson -wins-5g-network-contract-south-africa.

Trump, Donald J. 2018. "Presidential Executive Order on Streamlining and Expediting Requests to Locate Broadband Facilities in Rural America." The White House. https://www.whitehouse.gov/presidential-actions/presidential-executive-order -streamlining-expediting-requests-locate-broadband-facilities-rural-america/.

Venktess, Kyle. 2016. *Cwele Wins Spectrum Court Battle against Icasa.* Pretoria: fin24, September 30. https://www.news24.com/fin24/Tech/News/breaking-cwele-wins -spectrum-court-battle-against-icasa-20160930.

Vodacom. 2020. "Vodacom to Spend Over R500 Million within Two Months to Add Network Capacity during National State of Disaster." *Vodacom,* April 15. https:// www.vodacom.com/news-article.php?articleID=7476.

Waring, Joseph. 2019. "Korea Hits 3M 5G Subs as Base Stations Double." *Mobile World Live.* https://www.mobileworldlive.com/featured-content/top-three/south -korea-hits-3m-5g-subs-as-base-stations-double/.

Waring, Joseph. 2020. "Singtel, StarHub-M1 Combo Win Singapore 5G Licenses." *Mobile World Live,* April 30. https://www.mobileworldlive.com/asia/asia-news /singtel-starhub-m1-combo-win-singapore-5g-licences/.

Waterson, Jim, and Alex Hern. 2020. "At Least 20 UK Phone Masts Vandalised Over False 5G Coronavirus Claims." *The Guardian,* April 6. https://www.theguardian .com/technology/2020/apr/06/at-least-20-uk-phone-masts-vandalised-over-false -5g-coronavirus-claims.

WIA. 2016. *The Role of Street Furniture in Expanding Mobile Broadband.* Wireless Infrastructure Association. https://wia.org/resource-library/the-role-of-street -furniture-in-expanding-mobile-broadband/.

Xinhuanet. 2019. 曾剑秋：5G是国家信息化战略的延伸 会创造出更多需求 [*Zeng Jianqiu: 5G Is an Extension of the National Informatization Strategy that Will Create More Demand*], edited by Deng Cong. http://www.xinhuanet.com/info/2019-04/06 /c_137954518.htm.

Yin, Dave. 2019. "Industrial 5G Is Latest Ingredient in 'Made in China 2025'." *Caixing Global.* https://www.caixinglobal.com/2019-11-26/industrial-5g-is-latest -ingredient-in-made-in-china-2025-101487123.html.

Yuan, Eric S. 2020. *A Message to Our Users.* Zoom, April 1. https://blog.zoom.us/a -message-to-our-users/.

Zhou, Paul. 2019. "Why Are Governments Around the World Subsidizing 5G?" *LightReading.* https://www.lightreading.com/partner-perspectives-(sponsored -content)/why-are-governments-around-the-world-subsidizing-5g/a/d-id/754298.

Appendixes

APPENDIX A
Selected International Standards for Assessing and Limiting Human Exposure to Radio Frequency Emissions

TABLE A.1 **Overview of selected organizations and standards**

Organization/standard number or title	Description
CENELEC/EN 50360:2017/prA1 Product standard to demonstrate the compliance of wireless communication devices, with the basic restrictions and exposure limit values related to human exposure to electromagnetic fields in the frequency range from 300 MHz to 6 GHz devices used next to the ear	This product standard applies to wireless communication devices used near the human ear (for example, mobile phones, wireless headsets).
CENELEC/EN 50385:2017 Product standard to demonstrate the compliance of base station equipment with radiofrequency electromagnetic field exposure limits (110 MHz to 100 GHz), when placed on the market	This product standard is related to human exposure to radiofrequency electromagnetic fields transmitted by base station equipment in the frequency range 110 MHz to 100 GHz. The object is to assess the compliance of such equipment with the general public basic restrictions.
CENELEC/EN 50665:2017 Generic standard for assessment of electronic and electrical equipment related to human exposure restrictions for electromagnetic fields (0 Hz to 300 GHz)	The object of this generic standard is to provide a route for evaluation of such equipment against limits on human exposure to electric, magnetic, and electromagnetic fields, as well as induced and contact current.
ICNIRP/Guidelines for Limiting Exposure to Electromagnetic Fields (100 KHz to 300 GHz) March 2020	These guidelines are for the protection of humans exposed to radiofrequency electromagnetic fields in the range 100 KHz to 300 GHz.
IEC/62209-1:2016 Measurement procedure for the assessment of specific absorption rate of human exposure to radio frequency fields from handheld and body-mounted wireless communication devices—Part 1: Devices used next to the ear (frequency range from, 300 MHz to 6 GHz)	This guideline specifies protocols and test procedures for measurement of the peak spatial-average specific absorption rate induced inside a simplified model of the head with defined reproducibility. It applies to certain electromagnetic field transmitting devices that are positioned next to the ear, where the radiating structures of the device are near the human head, such as mobile phones, cordless phones, certain headsets, and so forth.

(continued)

TABLE A.1 *(continued)*

Organization/standard number or title	Description
IEC/62232:2017 Determination of radiofrequency field strength, power density, and specific absorption rate in the vicinity of radio base stations for the purpose of evaluating human exposure	This guideline provides methods for the determination of radiofrequency (RF) field strength and specific absorption rate in the vicinity of radio base stations (RBS) for the purpose of evaluating human exposure. This document • Considers intentionally radiating RBS that transmit on one or more antennas using one or more frequencies in the range of 110 MHz to 100 GHz • Considers the impact of ambient sources on RF exposure at least in the frequency range of 100 KHz to 300 GHz.
IEC/EN 62311:2019 Assessment of electronic and electrical equipment related to human exposure restrictions for electromagnetic fields (0 Hz to 300 GHz)	This document provides assessment methods and criteria to evaluate equipment against limits on exposure of people to electric, magnetic, and electromagnetic fields in the frequency range of 0 Hz to 300 GHz.
IEEE C95.1-2019 IEEE Standard for Safety Levels with Respect to Human Exposure to Electric, Magnetic, and Electromagnetic Fields, 0 Hz to 300 GHz	Safety limits for the protection of persons against the established adverse health effects of exposure to electric, magnetic, and electromagnetic fields in the frequency range of 0 Hz to 300 GHz are presented in this standard.
IEEE/IEC P63195-1 IEEE/IEC International Draft Standard—Measurement procedure for the assessment of power density of human exposure to radio frequency fields from wireless devices operating near the head and body— frequency range of 6 GHz to 300 GHz	The protocols and procedures apply to a significant majority of people, including children, during the use of handheld and body-worn wireless communication devices. These devices may feature single or multiple transmitters or antennas and may be operated with their radiating parts at distances up to 200 millimeters from a human head or body. This standard can be used to evaluate the power density compliance of different types of wireless communication devices used next to the ear, in front of the face, or mounted on the body.
ITU-T/Recommendation K.91 (June 2020) Guidance for assessment, evaluation, and monitoring of human exposure to radio frequency electromagnetic fields	This recommendation gives guidance on how to assess and monitor human exposure to radio frequency electromagnetic fields in areas with surrounding radiocommunication installations based on existing exposure and compliance standards in the 9 KHz to 300 GHz range.
ITU-T/Series K Supplement 9 5G technology and human exposure to radio frequency electromagnetic fields	This recommendation contains an analysis of the impact of the implementation of 5G mobile systems with respect to the exposure level of electromagnetic fields around radiocommunication infrastructure.

Source: Original table for this publication, with elaboration by the authoring team.
Note: CENELEC = European Committee for Electrotechnical Standardization; GHz = gigahertz; Hz = hertz; ICNIRP = International Commission on Non-Ionizing Radiation Protection; IEC = International Electrotechnical Commission; IEEE = Institute of Electrical and Electronics Engineers; ITU-T = International Telecommunication Union Telecommunications Standardization Sector; KHz = kilohertz; MHz = megahertz.

APPENDIX B
Selection of National Strategies Designed to Facilitate 5G-Enabling Environments

EXAMPLE 1: NATIONAL 5G STRATEGY IN THE UNITED KINGDOM (UK), THE REPUBLIC OF KOREA, AND PAKISTAN

The UK government created a Digital Infrastructure Officials Group under the Secretary of State for Culture, Media, and Sport. With a telecoms director as chair, the group is comprised of senior officials from various government departments to align projects that have digital infrastructure elements and to make sure other infrastructure projects are "future ready."

While the United Kingdom's leadership approach is at the macro-digital infrastructure level, the Republic of Korea takes a more 5G-specific approach with its "5G⁺ Strategy Committee." This committee has an overarching 5G deployment task force comprised of personnel from various ministries and the private sector. Korea's 5G⁺ Strategy Committee is comprised of about 10 ministries, while the Minister of Science and ICT and a private sector expert are co-chairs. This committee, which conjoins the interests of the public and private sectors, convenes monthly and quarterly 5G strategy review meetings and revises plans according to progress made and market developments. The committee also provides a platform for discussion, collects opinions, and handles complaints.

Among developing countries, Pakistan established the 5G Pakistan Plan Committee in early 2020, chaired by the Ministry of Information Technology and Telecommunication. The agenda for the committee's first meeting included discussions on key challenges of the existing 4G networks, spectrum availability and pricing, rights of way, 5G device ecosystem, and telecom taxation. Five subworking groups were formed: the National Broadband Plan, Spectrum, Regulations, Infrastructure, and Use Cases and Applications (Sahi 2020).

EXAMPLE 2: THE EUROPEAN UNION's 5G STRATEGY

The drafting of national 5G strategies in the European Union was triggered by the publication of "5G for Europe: An Action Plan." Its first action was "Encouraging Member States to develop, by end 2017, national 5G deployment roadmaps as part of the national broadband plans" (European Commission 2016). The 5G Working Group of the European Union's Communications Committee compiled and analyzed the member states' experiences and summarized the common elements and best practices to consider in national 5G strategies (COCOM 2018). Announcing the report, the head of the European Commission's Future Connectivity Systems unit commented, "The report shows that the two most critical issues appear to be spectrum assignment and facilitating the deployment of small cells" (Stuckmann 2018).

The best practices for developing national 5G strategies as recommended by the European Union's 5G Working Group (COCOM 2018) for the EU member states include the following:

- *Spectrum licensing.* Experimental licensing and authorization should be made available, including temporary spectrum rights to support research and experiments.

- *Promotion of testbeds.* National 5G testbeds and innovation platforms should be established and promoted.

- *Monitoring of trials and pilots.* Trials and pilots among industries should be monitored to "identify possible issues related to sectorial regulatory regimes that may constitute barriers for the take-up of new solutions" (COCOM 2018, 30).

- *Spectrum authorization models.* National spectrum plans should consider authorization models adapted to the characteristics of the different 5G bands and address the full range of options, including general authorization, individual rights, spectrum sharing, spectrum leasing or trading, and innovative or hybrid approaches.

- *Transparent spectrum plans.* Spectrum plans should be published, and spectrum availability should be made transparent to facilitate investment. Items to be included in these publications are the details of the authorization process, relevant award characteristics, timelines, and an outline of the decision-making process.

- *National roadmaps.* National 5G roadmaps should be established to drive the development of innovative use cases for 5G.

- *Public financing.* Public financing of current and future wireless and fixed broadband infrastructure is critical to complement private investments for 5G networks. The report further notes that public financing of backhaul infrastructure "should only take place where it is demonstrated that existing backhaul infrastructure is insufficient to cater for the needs of

5G networks and it should be carried out in such a way that it minimizes the distortion of competition with any already existing private networks" (COCOM 2018, 23).

EXAMPLE 3: KOREA's 5G+ STRATEGY

A forerunner in adopting the 5G network, the Korean government has emphatically noted its aim of pursuing 5G technologies not just to accelerate innovation in industry but also to improve quality of life (Government of the Republic of Korea 2019). 5G services were first launched in Korea in April 2019 following a strategic partnership between the government and the private sector, which was made possible through their mutual understanding of 5G as a key driver of the Fourth Industrial Revolution (Industry 4.0), along with big data and artificial intelligence.

The groundwork for this launch was established a few years earlier in November 2017, when the government announced the "I-Korea 4.0 Plan" to promote 5G in earnest (Government of the Republic of Korea 2017). The I-Korea 4.0 Plan is a broader national strategy aimed at realizing Industry 4.0 based on innovation-led economic growth, safety, and inclusiveness. The plan focuses on investments in information and communications technology, hyper-connected intelligent infrastructure, and innovation of the national research and development system. The Korean government envisions 5G as playing a critical role in achieving this revolutionary vision and aims to enhance the competitiveness of industries, the economy, and the entire nation through early commercialization of 5G.

As the I-Korea 4.0 Plan demonstrates, Korea's approach to 5G was to embed it tactically within a broader national strategy for realizing Industry 4.0. The Korean government subsequently established its 5G strategy and established 10 core industries and 5 core services to reap the full benefits of 5G technology. The 10 core industries include edge computing, information security, 5G vehicle-to-anything, connected robots, future drones, intelligent closed-circuit television, wearable devices, virtual and augmented reality, next-generation smartphones, and network equipment. The 5 core services include immersive content, smart factories, autonomous vehicles, smart cities, and digital health care.

The key themes from Korea's 5G strategy include the following:

- *Recognizing the potential of 5G* not just as a telecommunications network but also as an infrastructure for the entire national economy in the era of Industry 4.0.

- *Integrating 5G within the national agenda* to enable the nation's top leaders to directly exercise leadership around 5G and broader economic goals.

- *Establishing a cross-ministerial task force* to remove regulatory barriers to 5G deployment through a cross-ministerial response system.

- *Leveraging the public-private partnership model* to enable government, public sector, academia, and industry players to participate actively in 5G committees and planning forums.

- *Minimizing risks for the private sector* to encourage private sector investment in 5G.

- *Maximizing the long-term socioeconomic impact* through the design and execution of spectrum auctions and focus on retrieving appropriate spectrum fees only instead of maximizing fiscal revenue from fees.

The case of Korea demonstrates how critical an active partnership between the government and the private sector is in commercializing 5G networks, and it underscores the need for a holistic government promotion system.

EXAMPLE 4: THE US 5G FAST PLAN

Another early adopter of 5G networks and among the first to launch commercial services is the United States. In 2018, the US Federal Communications Commission (FCC) devised a three-pronged strategy, Facilitate America's Superiority in 5G Technology (the 5G FAST Plan), in a bid to gain global competitiveness through the government's commitment to free up more spectrum, review existing infrastructure policy that may act as a barrier to 5G deployment, and "modernize outdated regulations" to promote digital opportunities.

The US approach to 5G is more market led than those of the Asian front-runners, although it is not completely laissez-faire. Underscoring this faith in a market-based approach, the White House's statement on a National Strategy to Secure 5G focuses on driving the private sector–led domestic deployment of 5G, noting that the US government will "work with the private sector, academia, and international government partners to adopt policies, standards, guidelines, and procurement strategies that reinforce 5G vendor diversity to foster market competition" (White House 2020, 6).

Responding to the White House's 5G FAST Plan that focuses on business environment reform to accelerate the deployment of 5G, the US Congressional Research Service questioned whether the United States might fall behind other countries by taking on the same market-based strategy it had for previous generations of cellular technology. The US Congressional Research Service suggests that Congress may want to include more centralized government planning when it comes to 5G, noting that Congress should "monitor US progress on 5G deployment and technologies, consider whether there is a need for more planning and coordination with industry, and assess whether additional government involvement would help or hinder the efforts of US companies in the global race to 5G" (Gallagher and DeVine 2019, 31).

In February 2018 the FCC (2018) established a Strategic Plan for 2018–22, which states that it will do the following:

- *Maximize investment and promote light-touch regulation to facilitate 5G development* by devising rules that promote a light-touch regulation, facilities-based competition, flexible use policy, and freeing up of spectrum to encourage and facilitate the development of 5G networks.

- *Promote investment in infrastructure and 5G networks* by eliminating unnecessary administrative burdens to investment.

- *Remove regulatory barriers to broadband deployment.*

EXAMPLE 5: VIET NAM's NATIONAL 5G STRATEGY

Among developing economies, Viet Nam was an early adopter of a National 5G Strategy. Viet Nam views harnessing innovative technologies such as 5G as an important strategic task that can also aid labor productivity and economic competitiveness. With ambitious targets for 2025, 2030, and vision to 2045 etched in Resolution 52-NQ/TW of 2019, which outlines the guidelines and policies to actively participate in Industry 4.0, the Ministry of Planning and Investment (2019) underscores the urgency of the need to deploy 5G throughout the country and enable citizens' access to low-cost broadband internet to accelerate socioeconomic development.

TABLE B.1 **Additional examples of national 5G strategy documents in developing countries**

Country	Document title	Link
Brazil	Estratégia Brasileira de Redes de Quinta Geração (5G) [Brazilian Strategy for Fifth-Generation Networks]	https://www.mctic.gov.br/mctic/export/sites/institucional /sessaoPublica/arquivos/estrategia5g/Documento-base-da -Estrategia-Brasileira-de-5G.pdf
Colombia	Plan 5G Colombia	https://www.mintic.gov.co/portal/604/articles-118058 _plan_5g_2019120.pdf
Georgia	5G განვითარების ხელშეწყობის სტრატეგია [5G Development Promotion Strategy]	https://comcom.ge/uploads/other/3/3939.pdf
Malaysia	National 5G Task Force Report	https://www.mcmc.gov.my/skmmgovmy/media/General/pdf /The-National-5G-Task-Force-Report.pdf
Philippines	Philippine Roadmap for 5G Technology	https://ictecosystem.org.ph/wp-content/uploads/2019/05 /Project-Profile_5G_fy2019.pdf
Russian Federation	Концепция создания и развития сетей 5G/IMT-2020 в Российской Федерации [Concept for the creation and development of 5G / IMT-2020 networks]	https://digital.gov.ru/uploaded/files/kontseptsiya-sozdaniya -i-razvitiya-setej-5g-imt-2020.pdf

Note: 5G = fifth-generation mobile network technologies; IMT = international mobile telecommunications.

Viet Nam's objective in embracing Industry 4.0 is to advance development through a new growth trajectory and a robust digital economy underpinned by science and technology, innovation, and a high-quality workforce. As with other countries, emphasis is placed on the improvement of the quality of life and welfare of the people.

As the resolution notes, Viet Nam plans to cover 100 percent of the country with 5G networks by 2030. The country has even more ambitious plans for 2045, aspiring to become "one of the leading production and service centers, start-up and innovation centers of Asia's leading group; have high labor productivity, be capable of mastering and applying modern technology in all fields of socioeconomic, environment, defense and security" (Government of Vietnam 2019). The resolution also outlines the various aspects that the government will address to meet these targets, including the following:

- *Recognizing the true potential of 5G* by raising awareness at the government level and urgently and actively tackling the need to participate in Industry 4.0.

- *Building and developing a national innovation center* by focusing on the core technologies of Industry 4.0 and those with high levels of readiness, such as information technology, electronics, telecommunications, network safety and security, intelligent manufacturing, finance and banking, e-commerce, digital agriculture, digital travel, the digital culture industry, medicine, and education and training. This focus includes the following:

  - *Improving the efficiency of public investments in research and development* through adopting best practices in management.

  - *Promulgating a system of national standards and regulation* for the application and development of these core technologies.

  - *Utilizing existing high-tech parks and testing zones* to develop creative start-up ecosystems and enterprises.

- *Identifying and supporting priority industries and technologies* by creating a favorable business environment, supporting infrastructure investment, elevating human capital, and addressing public procurement practices.

- *Actively participating in regional and global legal frameworks* to develop horizontal policies that enable the digital economy, including laws on data, cybersecurity, digital identification, and authentication.

- *Mobilizing resources to invest in scientific research, innovation, and technology* by facilitating foreign investment to provide capital funding to innovative start-ups.

BIBLIOGRAPHY

COCOM (Communications Committee). 2018. "Report on the Exchange of Best Practices Concerning National Broadband Strategies and 5G 'Path-to-Deployment'." Working Document COCOM18-06REV-2, COCOM, Working Group on 5G, European Commission, Brussels, Belgium. https://ec.europa.eu/newsroom/dae/document .cfm?doc_id=55605.

European Commission. 2016. "5G for Europe: An Action Plan. COM(2016) 588 Final." European Commission, Brussels, Belgium. https://ec.europa.eu/transparency /regdoc/rep/1/2016/EN/1-2016-588-EN-F1-1.PDF.

Gallagher, Jill C., and Michael E. DeVine. 2019. "Fifth-Generation (5G) Telecommunications Technologies: Issues for Congress." CRS Report R45485, Congressional Research Service, Washington, DC. http://www.hsdl.org/?view &did=821493.

Government of the Republic of Korea. 2017. *People-Centered Plan for the Fourth Industrial Revolution to Promote Innovative Growth (I-KOREA 4.0).* Government of the Republic of Korea, Seoul. https://www.4th-ir.go.kr/article/detail/220.

Government of the Republic of Korea. 2019. *5G+ Strategy: To Realize Innovative Growth.* Government of the Republic of Korea, Seoul. https://www.msit.go.kr/cms/english /pl/policies2/__icsFiles/afieldfile/2020/01/20/5G%20plus%20Strategy%20to%20 Realize%20Innovative%20Growth.pdf.

Government of Vietnam, Ministry of Planning and Investment. 2019. "Resolution No. 52-NQ/TW on Guidelines and Policies to Actively Participate in the Fourth Industrial Revolution." Government of Vietnam, Hanoi, September 27. https:// thuvienphapluat.vn/van-ban/Dau-tu/Nghi-quyet-52-NQ-TW-2019-chinh-sach -chu-dong-tham-gia-cuoc-Cach-mang-cong-nghiep-lan-thu-tu-425113.aspx.

Sahi, Nokhaiz. 2020. "Ministry Holds 1st Meeting of Advisory Committee for 5G Planning in Pakistan." *The Nation*, May 10. https://nation.com.pk/10-Mar-2020/ministry-holds -1st-meeting-of-advisory-committee-for-5g-planning-in-pakistan.

Stuckmann, Peter. 2018. "Proud to Announce Our New Checklist for National 5G Strategies." *Shaping Europe's Digital Future Blog.* https://ec.europa.eu/digital-single -market/en/blogposts/proud-announce-our-new-checklist-national-5g-strategies.

US Federal Communications Commission. 2018. "Strategic Plan 2018–2022." FCC, Washington, DC. https://docs.fcc.gov/public/attachments/DOC-349143A1.pdf.

US White House. 2020. "National Strategy to Secure 5G." US White House, Washington, DC. https://www.hsdl.org/c/view?docid=835776.